AA008642

2nd Workshop on Theory, Modelling and Computational Methods for Semiconductors 2010

Journal of Physics: Conference Series Volume 242

York, United Kingdom
13-15 January 2010

Editors:

Max Migliorato **Matt Probert**

ISBN: 978-1-61738-801-9
ISSN: 1742-6588

Printed from e-media with permission by:

Curran Associates, Inc.
57 Morehouse Lane
Red Hook, NY 12571

Some format issues inherent in the e-media version may also appear in this print version.

Copyright© (2010) by the Institute of Physics
All rights reserved.

Printed by Curran Associates, Inc. (2010)

For permission requests, please contact the Institute of Physics
at the address below.

Institute of Physics
Dirac House, Temple Back
Bristol BS1 6BE UK

Phone: 44 1 17 929 7481
Fax: 44 1 17 920 0979

techtracking@iop.org

Additional copies of this publication are available from:

Curran Associates, Inc.
57 Morehouse Lane
Red Hook, NY 12571 USA
Phone: 845-758-0400
Fax: 845-758-2634
Email: curran@proceedings.com
Web: www.proceedings.com

TABLE OF CONTENTS

These conference proceedings contain the written papers of the contributions presented at the 2nd International Conference on: Theory, Modelling and Computational methods for Semiconductors. The conference was held at the St Williams College, York, UK on 13th–15th Jan 2010. The previous conference in this series took place in 2008 at the University of Manchester, UK.

The scope of this conference embraces modelling, theory and the use of sophisticated computational tools in Semiconductor science and technology, where there is a substantial potential for time saving in R&D. The development of high speed computer architectures is finally allowing the routine use of accurate methods for calculating the structural, thermodynamic, vibrational and electronic properties of semiconductors and their heterostructures. This workshop ran for three days, with the objective of bringing together UK and international leading experts in the field of theory of group IV, III-V and II-VI semiconductors together with postdocs and students in the early stages of their careers. The first day focused on providing an introduction and overview of this vast field, aimed particularly at students at this influential point in their careers.

We would like to thank all participants for their contribution to the conference programme and these proceedings. We would also like to acknowledge the financial support from the Institute of Physics (Computational Physics group and Semiconductor Physics group), the UK Car-Parrinello Consortium, Accelrys (distributors of Materials Studio) and Quantumwise (distributors of Atomistix).

The Editors

Acknowledgements

Conference Organising Committee: Dr Matt Probert (University of York) and Dr Max Migliorato (University of Manchester)

Programme Committee: Dr Marco Califano (University of Leeds), Dr Jacob Gavartin (Accelrys Ltd, Cambridge), Dr Stanko Tomic (STFC Daresbury Laboratory), Dr Gabi Slavcheva (Imperial College London)

Proceedings edited and compiled by Dr Max Migliorato and Dr Matt Probert

All papers published in this volume of *Journal of Physics: Conference Series* have been peer reviewed through processes administered by the proceedings Editors. Reviews were conducted by expert referees to the professional and scientific standards expected of a proceedings journal published by IOP Publishing.

2nd Workshop on Theory, Modelling and Computational Methods for Semiconductors IOP Publishing
Journal of Physics: Conference Series **242** (2010) 012001 doi:10.1088/1742-6596/242/1/012001

Calculating Strain Using Atomistic Simulations: a Review

V. Haxha and M.A. Migliorato

School of Electrical and Electronic Engineering, University of Manchester, Sackville Street, Manchester M13 9PL United Kingdom

max.migliorato@manchester.ac.uk

Abstract. We present a short review of methods of evaluating of strain from atomistic models in the context of linear elasticity.

1. Introduction

The use of strain to tune semiconductor device performance has been an important area of research since 1970. It is well known that it is possible to grow a semiconductor overlayer on the top of a thick layer substrate even if the lattice constants of the two layers are different.

A large strain can be then built into the overlayer, while still maintaining a long range atomic ordering in registry with the substrate monolayers. In the late 1940s Frank and Van der Merwe[1] provided essential understanding of this pseudo-morphic growth in their classic papers and they proposed a theory of epitaxy where layer growth occurs with lattices relaxed by misfit on the substrate. According to experimental work, so long as the lattice mismatch is below roughly 10% it is possible to grow an epitaxial film which is in complete cohesion with the lattice substrate;[2] the largest thickness achievable is called the critical thickness. Beyond the critical thickness there will be dislocations and as a result the atomic planes in the growth direction won't accurately align.

Strain is the result of a relatively small displacement of the atoms in the crystal lattice. The strain affects the materials properties of the semiconductor, like e.g. the lattice constant, energy gap and effective masses of electrons and holes.[2]

In the next sections we will review some of the basic concepts of elasticity in cubic crystals and further we will show how atomistic approaches can be used to extract the strain tensor directly from the atomic bonds.

2. Elasticity

For a cubic crystal it is possible to describe the strain (i.e. the ratio between the change as a result of stress and its original state) in the system by imaging a set of orthonormal vectors $\hat{x}, \hat{y}, \hat{z}$ in the unstrained system. As a result of the influence of a uniform deformation the axes will be distorted to x', y', z'. The new axes can be related to the previous ones by:

$$
\begin{aligned}
x' &= (1+\varepsilon_{xx})\hat{x} + \varepsilon_{xy}\hat{y} + \varepsilon_{xz}\hat{z} \\
y' &= \varepsilon_{yx}\hat{x} + (1+\varepsilon_{yy})\hat{y} + \varepsilon_{yz}\hat{z} \\
z' &= \varepsilon_{zx}\hat{x} + \varepsilon_{zy}\hat{y} + (1+\varepsilon_{zz})\hat{z}
\end{aligned}
\tag{1}
$$

© 2010 IOP Publishing Ltd

Journal of Physics: Conference Series **242** (2010) 012001 doi:10.1088/1742-6596/242/1/012001

The deformation in the system is defined by the coefficients $\varepsilon_{\alpha\beta}$.[2]

The new axes which have been created are not orthogonal in general. At the point r the effect of the deformation in the unstrained coordinates is:

$$\vec{r} = x\hat{x} + y\hat{y} + z\hat{z} \tag{2}$$

Then the new vector after distortion is:

$$\vec{r}' = x\vec{x}' + y\vec{y}' + z\vec{z}' \tag{3}$$

Note that the coefficients x,y, and z are unchanged while the new strained basis vectors are no longer necessarily unitary. The displacement caused by the deformation can be found from:

$$\vec{R} = \vec{r}' - \vec{r} = x(\vec{x}' - \hat{x}) + y(\vec{y}' - \hat{y}) + z(\vec{z}' - \hat{y})$$

$$\vec{R} = (x\varepsilon_{xx} + y\varepsilon_{yx} + z\varepsilon_{zx})\hat{x} + (x\varepsilon_{xy} + y\varepsilon_{yy} + z\varepsilon_{zy})\hat{y} + (x\varepsilon_{xz} + y\varepsilon_{yz} + z\varepsilon_{zz})\hat{z} \tag{4}$$

This can be written by defining the quantities u, v, and w :

$$\vec{R}(r) = u(r)\hat{x} + v(r)\hat{y} + w(r)\hat{z} \tag{5}$$

In the small strain limit:

$$x\varepsilon_{xx} = x\frac{\partial u}{\partial x}; y\varepsilon_{yy} = y\frac{\partial v}{\partial y}; z\varepsilon_{zz} = z\frac{\partial w}{\partial z} \tag{6}$$

Instead of using the 9 distortion components ε_{ij} to describe deformation, the strain components e_{ij} are conventionally used, defined by:

$$e_{xx} = \varepsilon_{xx} = \frac{\partial u}{\partial x}; e_{yy} = \varepsilon_{yy} = \frac{\partial v}{\partial y}; e_{zz} = \varepsilon_{zz} = \frac{\partial w}{\partial z} \tag{7}$$

The off – diagonal terms describe the angular distortion due to strain and can be written as:

$$e_{xy} = \vec{x}' \cdot \vec{y}' \approx \varepsilon_{yx} + \varepsilon_{xy} = \frac{\partial u}{\partial y} + \frac{\partial v}{\partial x}$$

$$e_{yz} = \vec{y}' \cdot \vec{z}' \approx \varepsilon_{zy} + \varepsilon_{yz} = \frac{\partial v}{\partial z} + \frac{\partial w}{\partial y} \tag{8}$$

$$e_{zx} = \vec{z}' \cdot \vec{x}' \approx \varepsilon_{zx} + \varepsilon_{xz} = \frac{\partial u}{\partial z} + \frac{\partial w}{\partial x}$$

It is worth noting how there is often confusion about the difference between strain and distortion components. In general there are 9 independent distortion components but only 6 strain components. However it is possible to find a basis transformation where the distortion matrix becomes triangular (upper or lower). In that case the non-zero distortions are exactly equal to the corresponding strain components.

The diagonal components of the strain or distortion tensor are those responsible for compressive or tensile strain. All the other components instead generate shear pressure.

Conventionally technologists don't need to deal with the full complexity of the strain tensor. In fact two dimensional epitaxial growth usually introduces only two strains, one parallel to the growth plane and one orthogonal (perpendicular strain). This happens because crystals tend to roughly conserve their volume and hence a negative (positive) pressure in the growth plane results in a counteracting positive (negative pressure) in the direction perpendicular to the plane.

It is often possible to estimate such strains easily using elastic theory, provided the elastic constants of the crystal are known. We are not going to formally define the elastic constants here. A formal derivation involving the stress-strain relations can be found in textbooks.[2]

There are in general 36 elastic constants for a cubic crystal, but these can be reduced in real structures by utilizing symmetry arguments.[2] For zinc- blende such symmetries reduce the total number to only

3 independent constants: c_{11},c_{12} and c_{44}. In a wurtzite crystal instead symmetry is lower and hence there are 5 independent elastic constants: c_{11},c_{12},c_{13},c_{33} and c_{44}.

In the zinc-blende crystal the 3 elastic constants are linked to particular deformations that in practice are difficult to achieve independently from each other. Deformations that are more practical are those described by the Bulk Modulus (B) and Shear constant (C'). For a zinc-blende crystal, B and C' are given by[2]

$$B = \frac{1}{3}(c_{11} + 2c_{12}) \tag{9}$$

$$C' = \frac{1}{2}(c_{11} - c_{12}) \tag{10}$$

However both the Bulk modulus and Shear constant can be easily expressed in terms of the second derivatives of the total energy of the system,

$$B = \frac{1}{\Omega}\frac{d^2E}{dv^2}\bigg|_{E=E_{coh}} \tag{11}$$

$$C' = \frac{1}{\Omega}\frac{d^2E}{d\varepsilon^2}\bigg|_{E=E_{coh}} \tag{12}$$

Where Ω is the atomic volume which for zinc-blende or diamond is equal to $a^3/8$ (a is the lattice constant). The expressions in Eqs. 11 and 12 provide a link to evaluating B and C' from interatomic potentials that give the total energy E. For the Bulk Modulus, v expresses the uniform deformation (hydrostatic) of any vector R' in the crystal:

$$R' = v^{\frac{1}{3}}R \tag{13}$$

While the shear constant C' is defined through the non-uniform deformation:

$$\begin{aligned} R'_x &= R_x(1+\varepsilon) \\ R'_y &= R_y/(1+\varepsilon) \\ R'_z &= R_z \end{aligned} \tag{14}$$

Such strains are realistic and provide a meaningful way of relating 2 of the 3 elastic constants to practical deformations. However it is worth noting how these deformations so far only modify the lattice vectors length and not the angle between then.

The third elastic constant c_{44} is instead related to the deformations which alter the angles between the lattice vectors:

$$\begin{aligned} R'_x &= R_x + \gamma R_y \\ R'_y &= R_y \\ R'_z &= R_z \end{aligned} \tag{15}$$

So that the relevant second derivative of the total energy is:

$$c_{44} = \frac{1}{\Omega}\frac{d^2E}{d\gamma^2}\Big|_{E=E_{coh}} \quad (16)$$

This evaluation is not straightforward. When performing the calculation one needs to take into account the internal displacements which move the interpenetrating fcc sublattices in the diamond structure along the z axis, and hence minimise the energy for every given γ, with respect of this internal displacement. This distortion has a deep meaning.[3] In fact in a single tetrahedron with one atom in the centre of the coordinate system, a positive shear distortion would pull apart the two atoms along the [110] direction, while simultaneously pushing closer the other two atoms in the $[1\bar{1}0]$ direction. As a result of these movements the atom at the centre of the tetrahedron is under radial and angular forces, which resist changes to bond lengths and bond angles respectively. The displacement that minimises the total energy (typically a downwards shift orthogonal to the shear plane for a positive shear) is a consequence of a subtle balance among the two forces and results in different bond lengths within the same tetrahedron.

This elastic behaviour has an important and often overlooked consequence: the 6 components of the strain tensor alone cannot define all possible strain states of a crystal. For shear strains a further quantity must be known to define the amount of internal displacement.

Such quantity is well known[4] and usually referred to as Kleinman's internal displacement parameter ζ, which defines the distance among the two sublattices as being $\zeta\gamma/4$ of the lattice constant a.

For the wurtzite structure similar definitions exist for the Bulk Modulus, Shear constant and Kleinman parameter which can be found in the literature.[5,6]

3. Evaluating Strain from an Atomistic Structure

Strain is usually a concept better suited to describing deformations in cubic cells and often one thinks of the entire crystal being composed by a series of cubic unit cell blocks. However strain also clearly affects the atomic positions within those blocks. Here we try to merge the two pictures and explain how to extract the strain tensor from atomic positions. This is useful in order to extract information from the result of e.g. Molecular Dynamics (MD). From Figure 1 we notice that each atom of a strained or unstrained crystal is always inside a tetrahedron. Hence by comparing each tetrahedron in a strained structure with a chemically equivalent bulk tetrahedron one can relate strained and unstrained lattice vectors and derive the components of the strain tensor.

Irrespective of the crystal structure, provided the x, y and z are chosen so that the [111] direction for zinc-blende corresponds to the [0001] for wurtzite, such comparison results in a simple system of 6 linear equations to finally obtain the components of the deformation tensor and hence the strain tensor. In practice this is done in the following way: there are 4 vectors connecting the second nearest neighbours surrounding the atom inside the tetrahedron. As there are only three degrees of freedom, one can in principle chose any three out of the four vectors. Whatever the particular choice of vectors, the problem becomes that of solving the three matrix problems of the form:

$$\begin{pmatrix} 1+\varepsilon_{xx} & \varepsilon_{xy} & \varepsilon_{xz} \\ \varepsilon_{yx} & 1+\varepsilon_{yy} & \varepsilon_{yz} \\ \varepsilon_{zx} & \varepsilon_{zy} & 1+\varepsilon_{zz} \end{pmatrix}\begin{pmatrix} a\cdot x_i \\ a\cdot y_i \\ a\cdot z_i \end{pmatrix} = \begin{pmatrix} a\cdot x_i' \\ a\cdot y_i' \\ a\cdot z_i' \end{pmatrix} \qquad i=1,2,3 \quad (17)$$

following the notation with and without the superscript to indicate strained and unstrained Cartesian coordinates and reminding that a is the lattice constant of the bulk material. Once solved the components of the distortion tensor can be easily related to those of the elastic strain by reminding that $e_{ii} = \varepsilon_{ii}$ and that:

$$\begin{aligned} e_{xy} &= \vec{x}' \cdot \vec{y}' = \varepsilon_{yx}(1+\varepsilon_{xx}) + \varepsilon_{xy}(1+\varepsilon_{yy}) + \varepsilon_{xz} \cdot \varepsilon_{yz} \\ e_{xz} &= \vec{x}' \cdot \vec{z}' = \varepsilon_{zx}(1+\varepsilon_{xx}) + \varepsilon_{xz}(1+\varepsilon_{zz}) + \varepsilon_{xy} \cdot \varepsilon_{yz} \\ e_{yz} &= \vec{y}' \cdot \vec{z}' = \varepsilon_{zy}(1+\varepsilon_{yy}) + \varepsilon_{yz}(1+\varepsilon_{zz}) + \varepsilon_{xy} \cdot \varepsilon_{zx} \end{aligned} \tag{18}$$

The last expression is calculated without assuming the small strain limit as in Eq. 8. The difference is often small but since these operations are usually automatically computed there is no need to introduce any approximations.

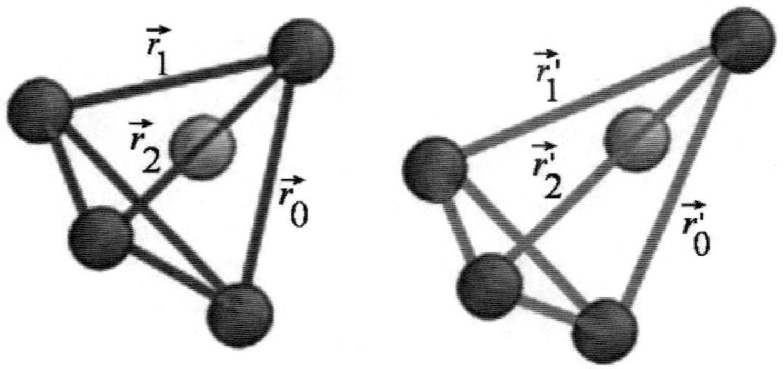

Figure 1: The unstrained and strained tetrahedron.

As explained earlier these set of operations fulfill the task of obtaining the 6 components of the strain tensor but does not provide any information on the sublattice displacement. In fact the position of the atom inside the tetrahedron does not enter the calculation at all. This is because the sublattice displacement can only be determined after the strain components have been identified. To identify the sublattice displacement one needs to first of all determine the ideal position (after strain but before any displacement) of the central atom within the tetrahedron. Then upon comparing the actual position and the ideal position (as calculated by simply applying the deformation tensor to the unstrained case, without considering internal relaxation) one obtains the displacement δ**r**. Such displacement can then be related to the Kleinman[4] parameter, which for convenience should be expressed as a 3D vector:

$$\begin{aligned} \delta x &= \frac{a}{4}\frac{\zeta_x \cdot e_{yz}}{1+e_{xx}} \\ \delta y &= \frac{a}{4}\frac{\zeta_y \cdot e_{xz}}{1+e_{yy}} \\ \delta z &= \frac{a}{4}\frac{\zeta_z \cdot e_{xy}}{1+e_{zz}} \end{aligned} \tag{19}$$

The reason to express the Kleinman[4] parameter as a vector is that recent work suggested that there is a strong dependence on strain for this quantity and hence even the word parameter is misleading.[7,8,9]

4. Alloys

The method outlined in the previous section is general and applicable to both pure and alloyed semiconductors. In both cases however one needs to clarify what the reference point is. In fact there

are two ways of defining strain. The first, often used by experimentalists, particularly electron microscopists, is to relate the increased or decreased lattice spacing of the particular semiconductor material or alloy one is studying in respect of that of the substrate or matrix material. This is obviously useful to experimentalists since the background material provides the calibration for the measurement. Alternatively, as theorists tend to do instead, the strain status is thought to be the deformation from the unstrained bulk case of the same material. The strain in this case is directly the quantity that is needed as input to evaluate the bandstructure modifications through strain.
The former presents however an inherent difficulty. For a semiconductor alloy the reference lattice parameter needs to be that of its theoretical unstrained case, which is often evaluated from Vegard's law[10] as an interpolation between binary compounds, for a given composition. However such composition in some cases (e.g. diffused or graded interfaces) is a local property of the alloy. Since it is difficult to evaluate such local composition from the first nearest neighbours alone, it is generally suggested to choose a suitable sphere around one atom and to obtain the composition by atom counting over several rings of nearest neighbours. This will give a suitable estimate of the local composition.

5. Practical Aspects of Evaluating Strain

In order to evaluate the strain tensor of a system of atoms whose positions are determined by MD, the following sequence of steps is required for each atom *i*:

1) Determine the 4 nearest neighbours. This is sometimes part of the MD output, but if not then it can be easily evaluated by searching through the atoms list within a small radius (dependent on the lattice parameter of the material) of the atom *i*. This is a triangular problem (if *i* is a neighbour of another atom *j*, then the opposite is also true) hence the search can be implemented rather quickly by avoiding searching twice on the same atoms.

2) Determine the local composition, as described in section 4, and use it to evaluate the local unstrained bond length, in the framework of Vegard's law[10] (using also the local composition) in the case of alloys.

3) Use the nearest neighbour list obtained during the first step to identify the coordinates of the nearest neighbours and use together with the local unstrained bond length obtained during step 2 in conjunction with Eqs. 17-18. Solve the resulting set of equations to obtain the 6 components of the strain tensor.

4) With the distortion components determined during step 3, determine the ideal position of atom *i* in respect of the four nearest neighbours by applying the distortion tensor to an unstrained set of atoms belonging to a material where the composition and lattice constant are those obtained during step 2. The difference between the ideal position and that found from the MD output gives δx, δy and δz of Eq. 19, from which the Kleinman[4] parameter is readily obtained. This step is normally not needed in most cases. We have previously noted how it is an important quantity if a comprehensive atomistic description of piezoelectricity is required.[7,9]

We will now show a sample of such strain calculations for a series of rather complicated structures consisting of a GaAs substrate, a thin two dimensional InAs layer, a three dimensional quantum dot island, a capping formed by a non uniform InGaAsSb alloy (with different Sb content for the 3 models) and a further GaAs capping.
We will only show the diagonal components of the strain, though the full strain tensor was calculated, as the diagonal components are those that contain more information about the lattice deformation.

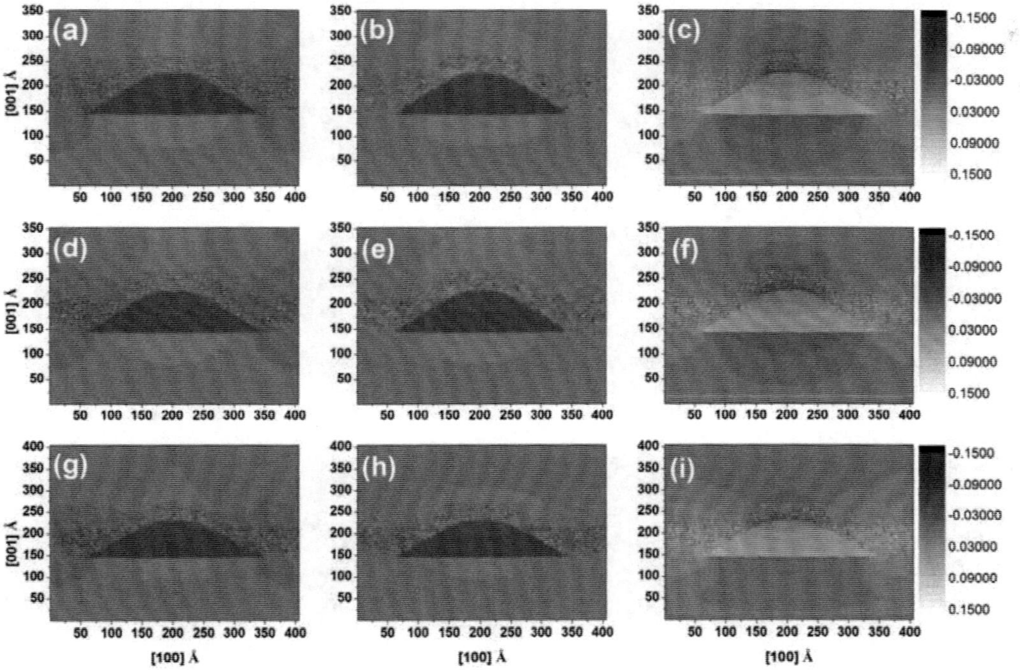

Figure 2: Strain maps for a InGaSbAs/InAs/GaAs Quantum Dot island. The InGaSbAs region is composed by In and Sb intermixed layers with a nominal Sb composition of 12% Sb ((a),(b) and (c)), 15% Sb ((d),(e) and (f)) and 20% Sb ((g),(h) and (i)). The first, second and third columns are the components of the strain ε_{xx} , ε_{yy} and ε_{zz}, respectively.[11]

What is noticeable from figure 2 is that by using atomistically determined strain maps it is possible to obtain smooth strain profiles in the regions where only binary semiconductor compounds are present. In the alloyed regions instead compositional disorder plays an important role and strain maps reproduce the coarse nature of the underlying material. In that respect this shows how, by preserving the nature of the material under study, atomistic modelling produces more realistic results compared to e.g. continuum elasticity theory.

6. Conclusions

So far we have described the mathematic involved in the evaluation of the strain from a relaxed atomistic model. Describing the force fields (e.g. bond order potentials[12], Keating's Valence Force Field[13] and Stillinger-Weber potential[14] that can be used to obtain the relaxed geometry is beyond the scope of this short review.

The methodology that we presented is general and can be used effectively and accurately in most atomistic level modelling of strain properties of semiconductor crystals.

References

[1] F.C. Frank , J. Van der Merwe, Proc. Roy. Sot. A, **198**, 205 (1949).

[2] J. Singh, *Physics of Semiconductors and Their Heterostructures*, McGraw – Hill, NY (1993).

[3] W.A. Harrison, *Electronic Structure and Properties of Solids*, Dover Publications, Inc., New York, 1989.

[4] L. Kleinman, Phys. Rev. **128**, 2614 (1962).

[5] A.F. Wright, J. Appl. Phys. **82**, 2833 (1997)

Journal of Physics: Conference Series **242** (2010) 012001 doi:10.1088/1742-6596/242/1/012001

[6] J.-M. Wagner and F. Bechstedt, Phys. Rev. B **66**, 115202 (2002)
[7] M.A. Migliorato, D. Powell, A.G. Cullis, T. Hammerschmidt and G.P. Srivastava, Phys. Rev. B **74**, 245332 (2006)
[8] D. Powell, M. A. Migliorato, A. G. Cullis, Phys. Rev. B **75**, 115202 (2007).
[9] R. Garg, A. Hüe, V. Haxha, M. A. Migliorato, T. Hammerschmidt, G. P. Srivastava, Appl Phys Lett **95**, 041912 (2009).
[10] L. Vegard, Z. Phys. **5**, 17 (1921).
[11] Reproduced from V. Haxha, I. Drouzas, J.M. Ulloa, M. Bozkurt, P.M. Koenraad, D.J. Mowbray, H.Y. Liu, M.J. Steer, M. Hopkinson, M.A. Migliorato, Phys Rev B **80**, 165334 (2009)
[12] J. Tersoff, Phys. Rev. Lett. **56**, 632 (1986); Phys. Rev.B **37**, 6991 (1988); Phys. Rev. B **39**, 5566 (1989).
[13] P.N. Keating, Phys. Rev. **145**, 637(1966).
[14] F. Stillinger, T. A. Weber, Phys. Rev. B **31**, 5262 (1985).

Tight-binding approach to calculation of localised perturbations in semiconductors

Eoin P. O'Reilly[1,2] and Andrew Lindsay[1]

[1]Tyndall National Institute, Lee Maltings, Cork, Ireland

[2]Department of Physics, University College Cork, Ireland

eoin.oreilly@tyndall.ie

Abstract. The replacement of As by N in GaAs introduces several pertubations, including both a change in potential at the N site and also long range strain relaxation in the crystal. It can be very useful to describe this perturbation in terms of the change in the Hamiltonian, ΔH, due to the introduction of N. Using plane-wave approaches, ΔH treats explicitly all atomic displacements in the structure. By contrast, a tight-binding approach allows a much simpler analysis. We illustrate the benefits of the tight-binding method by considering Ga(P,As:Bi) and GaNAs. We show that the tight-binding method provides a clear and quantitative explanation for many of the unusual electronic properties of these and related alloys, including the non-monotonic variation of electron effective mass, m_e, and gyromagnetic ratio, g_e, in GaNAs.

1. Introduction

There has been considerable interest over the last decade in the electronic structure of extreme semiconductor alloys, such as GaN_xAs_{1-x} and GaN_xP_{1-x}. It is well established that replacing a P or As atom by N introduces a defect level, which lies in the energy gap for GaP, and is a resonant state, above the conduction band edge in GaAs [1]. These localized states arise because of the large difference in size and electronegativity between N and the other group V elements. For conventional alloys such as $Al_xGa_{1-x}As$, the energy gap has a close to linear variation with composition x, and the electronic structure can be well described using the virtual crystal approximation, which assumes that the Al and Ga atoms can each be treated like an average (Al_xGa_{1-x}) atom. This approach fails for extreme alloys such as GaN_xAs_{1-x}, where the introduction of a small amount of N leads to a very rapid reduction in energy gap, with the energy gap decreasing by ~150 meV for x = 1%. Different models have been used to explain this observed reduction. The strong red shift can be modeled successfully by a band-anticrossing (BAC) interaction between the host III-V material conduction band edge (CBE) and a higher-lying band of N-related defect states which form above the CBE. This 2-level BAC model parameterizes and reproduces many of the key features of the alloy band structure, including the reduced energy gap [2], low electron mobilities [3], and a strong conduction band non-parabolicity [4]. Empirical pseudopotential calculations also describe well the key features of the band structure, and have been used to interpret the band structure in terms of the electronic states being a mixture of "perturbed host states" and N-related defect states [5].

We review here how calculations using the tight-binding model both support the BAC model and also provide useful additional insight into detail of the electronic structure of extreme alloys. We first

© 2010 IOP Publishing Ltd

describe how the localized basis states used in the tight-binding method make it a particularly useful approach to study perturbations in semiconductor alloys. We then summarise how it is considerably more straightforward to derive the BAC model using localised basis states, as in the tight-binding model, compared to the plane-wave basis states which are often used in empirical pseudopotential and in *ab-initio* approaches. We illustrate how the tight-binding method can quantitatively explain key experimental results for dilute nitride alloys, including the unexpectedly large and nonmonotonic variation of the electron effective mass observed in GaN_xAs_{1-x}. We finally present a preliminary analysis of the electronic structure of $GaP_{1-x}Bi_x$ and $GaAs_{1-x}Bi_x$, where a small proportion of P or As atoms are replaced by bismuth (Bi). Bi is also a group V element but is below Sb in the periodic table, and so is larger and more electropositive than P or As. We show that this leads to the formation of resonant defect levels in (above) the valence band for GaAsBi (GaPBi), which have a BAC-type interaction with the valence band maximum. Our calculations show that the observed reduction in energy gap with increasing Bi composition in $GaAs_{1-x}Bi_x$ includes contributions of similar magnitude from this valence-band BAC interaction and also from a downward shift of the CBE, which can be described using a conventional virtual crystal approach.

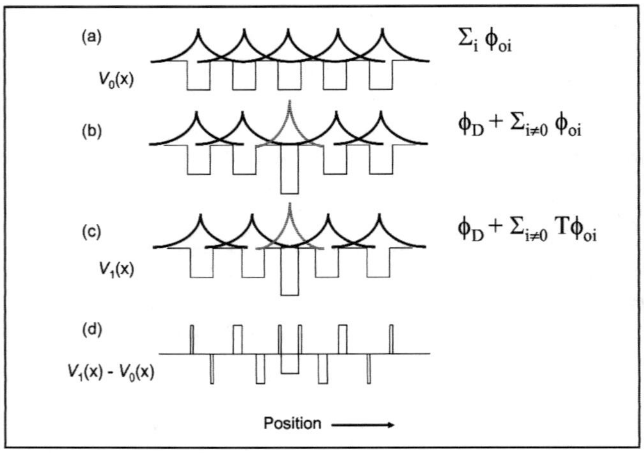

Figure 1: (a) Schematic of periodic potential, $V_0(x)$ and atomic basis states, ϕ_{0i}. (b) An impurity potential and basis state ϕ_D are introduced, with no change in period. (c) Long-range shift in atomic sites and in basis states, $T\phi_{0i}$, with (d) resultant long-range change in potential, $V_1(x) - V_0(x)$.

2. Tight-binding description of localized states

Replacing an As atom by N in GaAs has two main effects on the periodic crystal potential, as illustrated schematically in Fig. 1 (a). Firstly there is a change in the local potential at the N site (Fig. 1(b)) and secondly there is a change in the longer range potential, due to the crystal relaxing in response to the Ga-N bond length being 20% smaller than the Ga-As bond length (Fig. 1(c)). As a result the difference ΔH between the Hamiltonian H_0 for the perfect crystal and the Hamiltonian H_1 with N present includes not just a large local perturbation at the N site but also includes a significant long-range component, due to the crystal relaxation (Fig. 1(d)). If we use plane-wave basis states, then there is a significant contribution to the difference in the wavefunctions $|\psi_0\rangle$ and $|\psi_1\rangle$ in the perfect and relaxed crystals due to this change in the long-range potential. However, if we work with localized basis states which are centred on the atomic sites then it is much easier to analyse the impact of ΔH. We have for the unperturbed crystal in Fig. 1(a) that

$$H_0 \Sigma_i a^0_{ni} \phi_{0i} = E_{0n} \Sigma_i a^0_{ni} \phi_{0i}, \quad (1)$$

with the energy states found by solving the determinant equation $|H^0_{ij} - E\delta_{ij}| = 0$, with $H^0_{ij} = \langle\phi_{0i}|H_0|\phi_{0j}\rangle$, and where we assume that the orbitals $|\phi_{0i}\rangle$ and $|\phi_{0j}\rangle$ are orthogonal to each other. Turning to the perturbed crystal, we have, using the modified basis states $\phi_{10} = \phi_D$ and $\phi_{1i} = T\phi_{oi}$ ($i \neq 0$) from Fig. 1(c) that

$$H_1 \Sigma_i a^1_{ni} \phi_{1i} = E_{1n} \Sigma_i a^1_{ni}\phi_{1i} \quad (2)$$

with the energy states now found by solving the determinant equation $|H^1_{ij} - E\delta_{ij}| = 0$, where the matrix elements H^1_{ij} involve interactions between the modified basis states, ϕ_{1i}.

The BAC model explains the extreme band gap bowing in $GaInN_xAs_{1-x}$ in terms of an interaction between two levels, one at energy E_N associated with the localized N impurity states ψ_N, and the other at energy E_c associated with the extended CBE state ψ_{c0} of the GaInAs matrix, with the two states linked by a matrix element V_{Nc} describing the interaction between them [2]. The conduction band dispersion of $GaInN_xAs_{1-x}$ is then given by the lower eigenvalue E_- of the determinant

$$\begin{vmatrix} E_N & V_{Nc} \\ V_{Nc} & E_c + \dfrac{\hbar^2 k^2}{2m_c^*} \end{vmatrix} \quad (3)$$

where m^*_c is an appropriately chosen band edge effective mass for the Ga(In)As host matrix material.

To investigate the resonant state ψ_N, and its behaviour, we have developed an accurate sp^3s^* tight-binding (TB) Hamiltonian for $GaInN_xAs_{1-x}$ and related alloys [6]. This Hamiltonian fully accounts for the observed experimental data, and also gives results in good agreement with pseudopotential calculations [5,7]. By comparing the calculated CBE states found in large supercells ψ_{c1} (defined with respect to states equivalent to the set ϕ_{1i}) and ψ_{c0} (defined with respect to states equivalent to the set ϕ_{0i}), we can derive the nitrogen resonant state ψ_{N0} associated with an isolated N atom. The calculation effectively treats the basis states ϕ_{1i} and ϕ_{0i} as equivalent. With this assumption of equivalent bases, ψ_{c1} can then be calculated as a linear combination of ψ_{c0} and ψ_{N0}, with ψ_{N0} given by

$$\psi_{N0} = \frac{\psi_{c1} - \alpha\psi_{c0}}{\sqrt{1-\alpha^2}} \quad (4)$$

where $\alpha = \langle \psi_{c1}|\psi_{c0}\rangle$, and where we have set $\langle \phi_{1i}|\phi_{0j}\rangle = \delta_{ij}$ when treating the change in basis states between Figs. 1(a) and (c). With this assumption, we find that ψ_{N0} is highly localised, with over 50% of its probability density on the N site and the four neighbouring Ga atoms.

To investigate the role of disorder, we extended the tight-binding and two-level model to disordered GaN_xAs_{1-x} supercells. We find for $Ga_LN_MAs_{L-M}$ supercells containing ~1000 atoms that the energy and wavefunction of the CBE state can be well represented by considering the interactions between a **L**inear **C**ombination of **I**solated **N**itrogen resonant **S**tates and the GaAs CBE wavefunction, ψ_{c0} (LCINS method). For any given $Ga_LN_MAs_{L-M}$ supercell, we associate a localised resonant state ψ_{Ni} with each of the M nitrogen atoms ($i = 1,\ldots M$), and a wavefunction ψ_{c0} with the host matrix CBE. We then analyse the supercell conduction band states by solving the $(M+1) \times (M+1)$ matrix equation linking the M defect states and the CBE state, $H_{ij}\phi_j = E\, S_{ij}\phi_j$, where $H_{ij} = <\psi_{Ni}|H|\psi_{Nj}>$ and $H_{i,M+1} = <\psi_{Ni}|H|\psi_{c0}>$, $1 \le i,j \le M$, with H the full GaNAs Hamiltonian, and $S_{ij} = <\psi_{Ni}|\psi_{Nj}>$ a matrix reflecting that neighbouring N states can overlap each other.

3. LCINS Analysis of GaNAs band structure

The filled data points in Fig. 2 show the low temperature (a) electron effective mass m_e and (b) gyromagnetic ratio g_e determined by magnetophotoluminescence for a wide range of GaN_xAs_{1-x} samples [6,8]. The solid lines show the predicted variation of m_e and g_e in bulk GaN_xAs_{1-x} using the two-level BAC model of Eq. (3). This model significantly underestimates the measured mass, even for x as low as 0.1%. The open symbols show the low temperature values of m_e and g_e calculated for selected compositions, x, using the LCINS model and including resonant defect states associated with a full random distribution of N atoms.

The calculated and experimental data are in remarkable agreement with each other, confirming that hybridisation between the CB edge and nitrogen cluster states causes the observed enhancement of effective mass values. The mass m_e initially increases with composition x, as the band edge passes between $x = 0.1$ and 0.2% through N-N pair states, where a Ga atom has two N neighbours. For $x >$ 0.2%, we might expect a decrease in m^*_e towards the 2-level BAC value, as the CBE E_- level passes through the N pair states. However, the density of other low-energy N-related states starts to increase markedly with increasing x. These combined interactions lead to a further reduction in the Γ character at the band edge, consistent with the additional, sudden increase in m_e observed experimentally in Fig. 2 in this concentration range. Similar effects are also observed in the theoretical and experimental data for g_e. These results highlight the advantages of using a tight-binding approach with localised basis states to analyse the effects of localised perturbations in semiconductors.

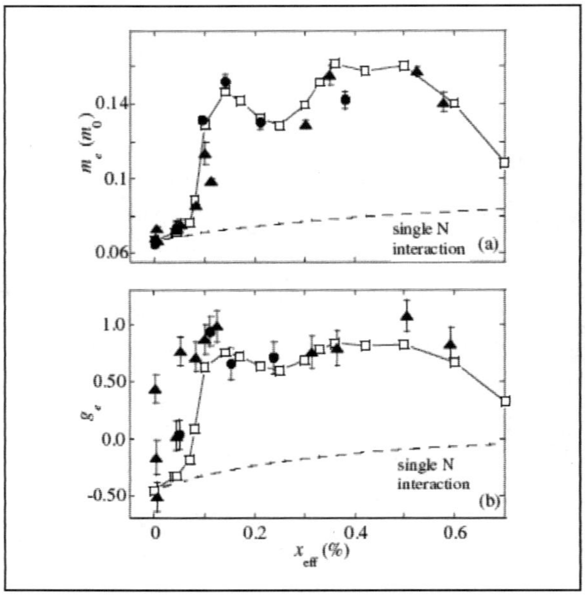

Figure 2: Dependence (*a*) of the electron mass m_e and (*b*) gyromagnetic ratio g_e on nitrogen effective concentration in GaN_xAs_{1-x} (solid symbols: experiment; open square symbols: full tight-binding-based theory). The dashed lines show the calculated values of m_e and g_e based on the BAC model of Eq. (3) [From ref. 6].

4. Band structure of GaAsBi and GaPBi

It has been proposed that replacing arsenic by bismuth (Bi) could introduce a similar BAC effect to that in GaNAs [9]. Bi is much larger than As, and is also less electronegative. In this case the resonant defect level should therefore lie in the valence band and the anticrossing interaction should occur between the Bi defect level and the valence band edge (VBE). The energy gap of $GaAs_{1-x}Bi_x$ is observed both theoretically [10] and experimentally [11] to decrease quite rapidly with increasing Bi composition, but it has not yet been confirmed as to whether this reduction is due to an anticrossing interaction [9] or is better explained using more conventional alloy models. To address this question, it is useful to first consider GaPBi, because Bi is sufficiently different to P that it introduces defect states just above the VBE in GaP [12]. Because these states are in the energy gap, it is then much easier to identify and follow the interaction between the VBE and these levels as a function of increasing Bi composition. Our calculations show clearly for GaPBi that there is an anticrossing interaction between the Bi defect states and the GaP VBE.

Full tight-binding calculations were carried out on 1000-atom supercells to compare the valence band structure of GaY and of $GaY_{1-x}Bi_x$ (Y = P, As) as a function of Bi composition x for Bi compositions ranging from 0.2% (1 Bi atom) to 7% (35 Bi atoms), with the Bi atoms placed at random on the group V sites. We first analyse the electronic structure of $GaP_{1-x}Bi_x$, where it is easier to follow the evolution of the Bi energy states.

When we replace a single P atom by Bi in a $Ga_{500}P_{500}$ supercell, we calculate that this gives a four-fold degenerate defect level above the GaP VBE. These four levels have the same symmetry as the heavy- and light-hole states at the valence band maximum. Even in a 1000 atom supercell, these defect levels interact with the host matrix VBE states. We calculate that the isolated Bi impurity states are .052 eV above the GaP VBE, with each impurity state 62% localised on the Bi atom.

Having established the character of the isolated Bi impurity states, we now investigate how the Bi levels evolve as we increase the number L of Bi atoms in a $Ga_{500}P_{500-L}Bi_L$ supercell. In order to trace the evolution of the highest valence states, we use the sp^3s^* Hamiltonian to calculate exactly the energy states and wavefunctions at the zone centre of the $Ga_{500}P_{500-L}Bi_L$ supercell. Figure 3 shows the calculated $Ga_{500}P_{500-L}Bi_L$ supercell energy spectrum projected onto the unperturbed GaP VBE heavy-hole and light-hole wavefunctions, $\psi_{v\alpha 0}$:

$$G_{\Gamma}(E) = \sum_{i,\alpha} \left|a_{\Gamma\alpha i}\right|^2 T\,(E - E_i) \tag{5}$$

where $a_{\Gamma\alpha i}$ is the amplitude of the ith eigenstate, of energy E_i projected onto the αth valence band zone centre state ($\alpha = 1,..\ 4$).

We see in the left hand panel of Fig. 3 for a single Bi atom in the supercell ($x = 0.2\%$) that the amplitude of the unperturbed GaP valence band edge states, $\psi_{v\alpha 0}$, is distributed over two levels, E_- and E_+, as expected from the BAC model of Eq. (3). Because of the T_d symmetry around the isolated Bi atom, the four E_--related levels are degenerate and so cannot be distinguished for $M = 1$ ($x = 0.2\%$). We calculate for the isolated Bi impurity that the band anticrossing interaction V_{MD} between the Bi defect levels E_{Bi} and the VBE E_v is of magnitude $V_{MD} = \beta x^{½}$, with $\beta = 1.67$ eV. The calculated magnitude of the interaction parameter β in $GaP_{1-x}Bi_x$ is then comparable to that which we have calculated previously for $GaP_{1-x}N_x$ ($\beta = 1.74$ eV) and for $GaAs_{1-x}N_x$ (($\beta = 2.00$ eV) [6].

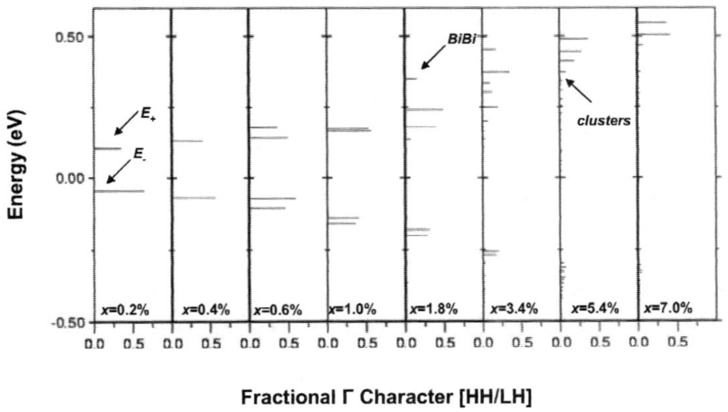

Figure 3: Calculated evolution of fractional Γ character $G_\Gamma(E)$ (defined in Eq. (5)) for a series of $Ga_{500}P_{500-L}Bi_L$ supercells. The BAC interaction can be clearly observed between Bi defect states in the energy gap and the valence band edge for $x \leq 1\%$. States due to Bi pairs and Bi clusters are also observed at higher Bi composition.

When we consider a single Bi atom in a $Ga_{500}As_{499}Bi_1$ supercell ($x = 0.2\%$), we find that the HH and LH Γ character remains predominantly on the highest valence states in this case, with a small fraction of the Γ character (< 3%) also associated with an energy state 0.25 eV below the GaAs valence band maximum. Using Eq. (4), we identify that the states which mix with the GaAs VBE states are localised Bi resonant defect levels, with very similar character to the localised defect states found in GaPBi. The Bi state is calculated to be 65% localised on the Bi atom in the $Ga_{500}As_{499}Bi_1$ supercell, compared to being 62% localized in the equivalent GaPBi supercell. We calculate by evaluating $V_{MD} = \langle\psi_{Bi}|H|\psi_{v\alpha 0}\rangle$ that the band anticrossing interaction V_{MD} between the Bi resonant defect levels E_{Bi} and the valence band edge E_v is of magnitude $V_{MD} = \beta x^{½}$, with $\beta = 1.22$ eV. The calculated magnitude of the interaction parameter β in $GaAs_{1-x}Bi_x$ is then smaller than the value in GaPBi, reflecting that there is less difference in electronegativity and size between a Bi and As atom than there is between Bi and P.

In addition to the BAC interaction pushing the VBE upwards, the conduction band edge energy is calculated to shift down by about 40 meV per % of Bi replacing As, but the fractional Γ character, $f_{\Gamma c} = |\langle\psi_c|\psi_{c0}\rangle|^2$ at the CBE remains close to 1. The evolution of the conduction band states is then best described using a conventional alloy model, including band gap bowing,

In summary, our calculations confirm that the highest valence band states in Ga(P,As)Bi can be described using a BAC model. The reduction in energy gap with composition in GaNAs is due predominantly to the conduction band BAC interaction, with most of the energy gap difference giving rise to a large conduction band offset and virtually zero valence band offset between GaAs and GaNAs. By contrast, the conduction band edge energy also varies with composition in GaAsBi, so that the reduction in energy gap with increasing Bi composition then contributes to approximately equal conduction and valence band offsets between GaAsBi and GaAs.

5. Conclusions

In summary, we have described how a well parameterised tight-binding method can be particularly suited to investigating the effects of localised perturbations in the electronic structure of semiconductor alloys. We illustrated the method by considering two examples. We first reviewed how tight-binding calculations provide a clear and quantitative explanation for many of the unusual electronic properties of GaNAs, including the non-monotonic variation of electron effective mass, m_e, and gyromagnetic ratio, g_e, in this alloy. We then presented for the first time an analysis which shows that the highest valence band states in Ga(P,As)Bi can be described using a BAC model, but that the conduction band edge energy also varies with composition in GaAsBi, so that the reduction in energy gap with increasing Bi composition then contributes to approximately equal conduction and valence band offsets between GaAsBi and GaAs.

Acknowledgements This work was supported by Science Foundation Ireland.

References

[1] For a review see: " *Physics and Applications of Dilute Nitrides*", ed. I. A. Buyanova and W. M. Chen (Taylor & Francis, New York, 2004); "*Dilute Nitride Semiconductors*", ed. M. Henini (Elsevier Science Ltd, 2005); "*Dilute III-V Nitride Semiconductor and Material Systems*", Springer Series in Materials Science, vol. **105**, ed. A. Erol (Springer 2008)

[2] W. Shan, W. Walukiewicz, J.W. Ager III, E.E. Haller, J.F. Geisz, D.J. Friedman, J.M. Olson and S.R. Kurtz, Phys. Rev. Lett. 82, 1221 (1999)

[3] S. Fahy and E.P. O'Reilly, Appl. Phys. Lett. **83**, 3731 (2003)

[4] J. Endicott, A. Patanè, J. Ibáñez, L. Eaves, M. Bissiri, M. Hopkinson, R. Airey, and G. Hill, Phys. Rev. Lett. **91**, 126802 (2003)

[5] P.R.C. Kent and A. Zunger, Phys.Rev.B **64**, 115208 (2001)

[6] E.P. O'Reilly, A. Lindsay, P.J. Klar, A. Polimeni and M. Capizzi, Semicond. Sci. Technol. **24**, 033001 (2009)

[7] T. Mattila, S.H. Wei and A. Zunger, Phys.Rev. B **60** R11245 (1999)

[8] F. Masia, G. Pettinari, A. Polimeni, M. Felici, A.Miriametro, M. Capizzi, A. Lindsay, S.B. Healy, E. P. O'Reilly, A. Cristofoli, G. Bais, M. Piccin, S. Rubini, F. Martelli, A. Franciosi, P. J. Klar, K. Volz, and W. Stolz, Phys. Rev. B **73**, 073201 (2006)

[9] K. Alberi, O.D. Dubon, W. Walukiewicz, K.M. Yu, K. Bertulis and A. Krotkus, Appl. Phys. Lett. **91**, 051909 (2007)

[10] A. Janotti, S.-H. Wei, S. B. Zhang, Phys. Rev. B **65**, 115203 (2002)

[11] K. Alberi, J. Wu, W. Walukiewicz, K. M. Yu, O. D. Dubon, S. P. Watkins, C. X. Wang, X. Liu, Y.-J. Cho and J. Furdyna, Phys. Rev. B **75**, 45203 (2007)

[12] F. A. Trumbore, M. Gershenzon, and D. G. Thomas, Appl. Phys. Lett. **9**, 4 (1966)

Journal of Physics: Conference Series **242** (2010) 012003 doi:10.1088/1742-6596/242/1/012003

First principles calculations of the scattering cross section of substitutional carbon in silicon

M.P. Vaughan[1] **and S. Fahy**[1,2]

[1]Tyndall National Institute, Lee Maltings, Cork, Ireland; [2]Department of Physics, University College Cork, Cork, Ireland

E-mail: martin.vaughan@tyndall.ie

Abstract. A method of calculating scattering cross-sections based on first principles electronic structure calculations, previously used for alloy scattering in SiGe is presented and applied to substitutional carbon in silicon. It is found that the intravalley scattering is the predominant scattering mechanism with a contribution to the scattering rate around a factor of 4 greater than intravalley alloy scattering in SiGe.

1. Introduction

The introduction of carbon into SiGe/Ge heterostructures is of technological interest due the dual effects of strain-compensation in the SiGe layers [1] and suppression of the out-diffusion of p-type acceptors, in particular boron [2] and indium [3], during wafer fabrication. Transport measurements of tensile strained $Si_{1-y}C_y$ and compressive strained $Si_{1-x-y}Ge_xC_y$ show a degradation of the n and p-type Hall mobility respectively as the carbon concentration is increased [4].

We investigate n-type scattering using a model of alloy scattering based on first principles electronic structure calculations [5]. The model has previously been used to calculate both n-type [5, 6] and p-type mobility [7] in SiGe alloys. The methodology is presented in section 2 and the details of the self-consistent calculations and extracted scattering parameters are given in section 3.

2. Theory

The method of calculating the scattering cross-section described here [5, 6] is based on *ab initio* electronic structure calculations in large supercells. Initially, a supercell is constructed for the unperturbed material, which, due to the periodic boundary conditions, will often have degenerate eigenvalues at points where high symmetry points in the primitive Brillouin zone (BZ) have been folded back into the supercell BZ. Next, a similar supercell is constructed with one of the original (in our case silicon) atoms replaced with a defect (substitutional carbon). This introduces a perturbing potential that causes the degenerate eigenvalues to split. The scattering cross-section is then proportional to the square of these splittings multiplied by the density of states.

The folding back of **k**-points occurs when the smaller reciprocal lattice vectors of the supercell map a **k**-point within the supercell BZ to another **k**-point outside of it but still within the primitive cell BZ. For instance, for a crystal with a face-centred cubic (FCC) primitive cell

© 2010 IOP Publishing Ltd

Journal of Physics: Conference Series **242** (2010) 012003 doi:10.1088/1742-6596/242/1/012003

such as silicon, a 64-atom supercell may be constructed with simple cubic (SC) symmetry with reciprocal lattice vectors (in Cartesian coordinates) $\mathbf{g}_1 = (2\pi/a)(1/2, 0, 0)$, $\mathbf{g}_2 = (2\pi/a)(0, 1/2, 0)$ and $\mathbf{g}_3 = (2\pi/a)(0, 0, 1/2)$, where a is the lattice constant. Each $\pm\mathbf{g}_i$ maps the Γ point to one of the 6 equivalent points halfway along the Δ line (which we shall denote by X/2) of the primitive BZ. Similarly, the vectors $\pm 2\mathbf{g}_i$ map the Γ point to the 6 X points. Since a translation of the supercell BZ by its reciprocal lattice vectors must leave the physics unchanged, these points must be folded back to the Γ point. Thus, when we carry out first principles calculations for the bandstructure of the pure material, we obtain two sets of 6-fold degenerate states at $\mathbf{k} = 0$ (as well as the original Γ states and other states folded back to this point).

Consider the ways in which carriers in the 6 equivalent Δ valleys may scatter due a perturbing potential $\Delta V = H - H_0$, where H and H_0 are the Hamiltonians of the perturbed and unperturbed system respectively. Denoting the unperturbed states by $|\psi_\beta\rangle$, where β labels the band, the scattering matrix in the Born approximation is $\langle\psi_\alpha|\,\Delta V\,|\psi_\beta\rangle$. When $\alpha = \beta$, we have intravalley scattering. For intervalley scattering, only two types of scattering are distinguishable by symmetry considerations: g-type, where the scattering is from one valley to its mirror image along the same Cartesian coordinate, and f-type, where the scattering is to a valley in an orthogonal direction. The matrix elements for g-type, f-type and intravalley scattering may be denoted respectively by V_g, V_f and V_Δ . Diagonalisation of the matrix yields one non-degenerate and 2 and 3-fold degenerate eigenvalues ΔE_1, ΔE_2 and ΔE_3 respectively. Writing the original scattering matrix elements in terms of these, we find [6]

$$V_\Delta = \frac{1}{6}\left(\Delta E_1 + 2\Delta E_2 + 3\Delta E_3\right), \tag{1}$$

$$V_g = \frac{1}{6}\left(\Delta E_1 + 2\Delta E_2 - 3\Delta E_3\right), \tag{2}$$

$$V_f = \frac{1}{6}\left(\Delta E_1 - \Delta E_2\right), \tag{3}$$

In the more general situation, where we go beyond the Born approximation using the results of supercell calculations, the construction of the scattering matrix with the exact states $|\phi_\beta\rangle$, requires that far from the impurity site $|\phi_\beta\rangle \rightarrow |\psi_\beta\rangle$. However, this condition cannot be satisfied for any eigenstate $|\phi_i\rangle$ within the confines of the supercell, so instead we construct $|\phi_\beta\rangle$ from the projection of the unperturbed states onto a subset of the $|\phi_i\rangle$ giving the greatest overlap according to $|\phi_\beta\rangle = \sum_{i=1}^{M} |\phi_i\rangle\langle\phi_i|\psi_\beta\rangle$. The scattering matrix element is then

$$\langle\psi_\alpha|\,\Delta V\,|\phi_\beta\rangle = \langle\psi_\alpha|\,\Delta V \sum_{i=1}^{M} |\phi_i\rangle\langle\phi_i|\psi_\beta\rangle = \sum_{i=1}^{M} \Delta E_i^\alpha \langle\psi_\alpha|\phi_i\rangle\langle\phi_i|\psi_\beta\rangle\,, \tag{4}$$

where $\Delta E_i^\alpha = E_i - E_\alpha$ is the difference between the perturbed and unperturbed eigenvalues E_i and E_α respectively. Putting $dV_\alpha^{red} = \sum_{i=1}^{M} \Delta E_i^\alpha |\phi_i\rangle\langle\phi_i|$, Eq. (4) may be written as

$$\langle\psi_\alpha|\,\Delta V\,|\phi_\beta\rangle = \langle\psi_\alpha|\,dV_\alpha^{red}\,|\psi_\beta\rangle\,. \tag{5}$$

For electron scattering, this matrix may then be constructed from, say, the n lowest lying conduction band states using Eq. (4), giving an $n \times n$ matrix. To calculate the scattering parameters given in Eqs (1) - (3), we must extract from this $n \times n$ matrix the 6×6 submatrices corresponding to the originally 6-fold degenerate states (for a 64-atom cell, the X and X/2 states).

When calculating the scattering due to a localised impurity, large supercells are required to ensure that the pertubed system has converged to the original system far from the impurity site and that the artificially imposed boundary conditions on the supercell are not giving rise

Journal of Physics: Conference Series **242** (2010) 012003 doi:10.1088/1742-6596/242/1/012003

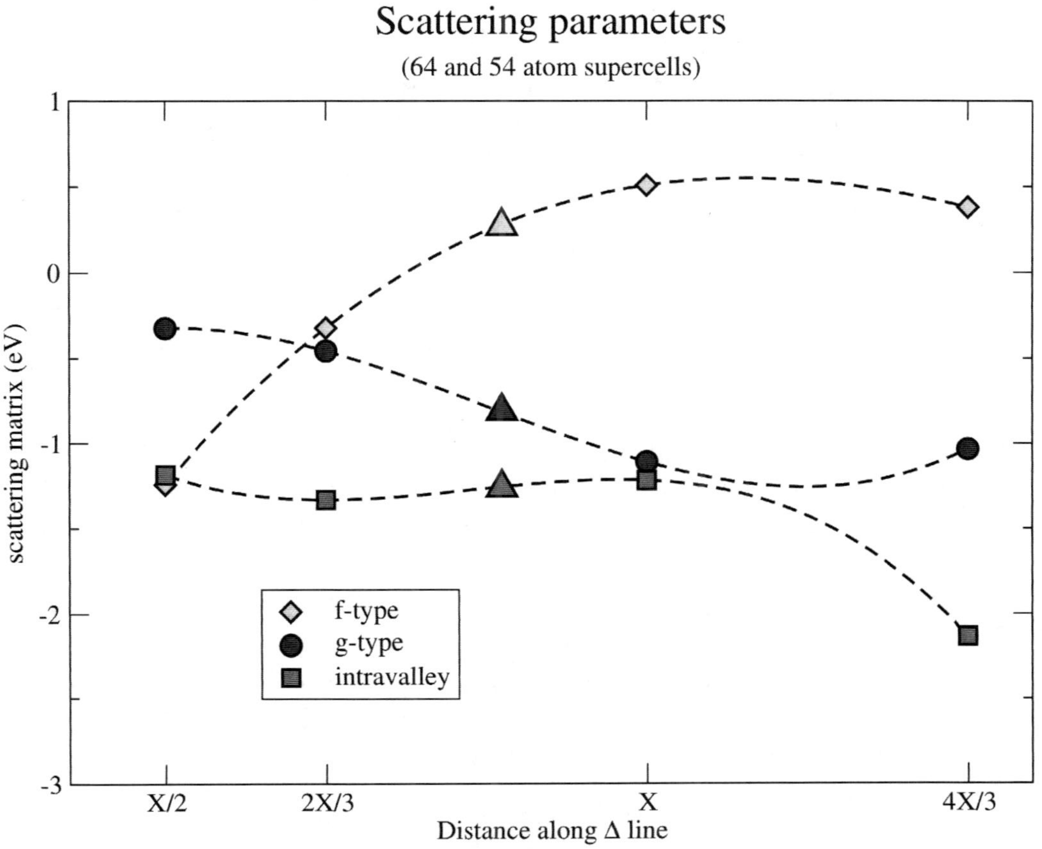

Figure 1. The parameters for intravalley and *f*-type and *g*-type intervalley scattering. The X/2 and X points are found from calculations of a 64-atom supercell calculations; the 2X/3 and 4X/3 points from a 54-atom supercell. The values at the Δ valley minimum (shown as triangles) are calculated from the polynomial interpolation of these points.

to spurious results. However, unless the impurity forms a bound state, the splitting of the degenerate energy levels will be inversely proportional to the supercell volume. Hence, the scattering parameters must be multiplied by the cell size, N, to give a consistent scattering matrix to be used in mobility calculations [5, 6].

Another important point that needs to be addressed is that the zero of the potential in a supercell is arbitrary. However, Eqs. (1) - (3) show that whilst intervalley scattering depends only on the relative shifts in the eigenvalues, intravalley scattering requires knowledge of the absolute values of the energy shifts. The approach used here is to take the atomic position of the Si atom furthest away from the repeated C atoms at the vertices of the supercell that remains fixed on ionic relaxation. For the 64-atom cubic supercell, this is the atom at the centre of the cell; for the 54 atom FCC supercell, this is the atom at $(4/9)a(1,1,1)$. The differences in the total potential of the pure Si supercell and relaxed Si(N-1)C(1) structure are then averaged over increasing radii from this atomic centre. The value of the averaged difference Where the derivative with respect to radius is zero is taken as the energy offset to correct the relaxed eigenvalues.

3. Results

3.1. Details of electronic structure calculations

The electronic structure calculations were performed using the ABINIT software [8] using the local density approximation (LDA) for the exchange and correlation functional. Troullier-Martins pseudopotentials [9] were used with an energy cutoff of 40 Hartree, chosen on the basis of convergence tests for 8-atom supercells. The ionic positions were relaxed using modified Broyden-Fletcher-Goldfarb-Shanno (BFGS) minimisation [10–12], which takes into account both the total energy as well as the energy gradients [13].

3.2. Scattering parameters

Calculations for 54 and 64 supercells were carried out and the scattering parameters (multiplied by cell size) extracted. Note that these values need to be squared and multiplied by the density of states to obtain the actual scattering cross-section via Fermi's Golden Rule. The results are plotted in Fig. 1 at different points along the Δ line between the Γ and X points. These points indicate the real variation of the scattering parameters with **k**-vector. Interpolating these data at the conduction band minimum in Si gives us the values , V_f = 0.27 eV, V_g = -0.81 eV and V_Δ = -1.26 eV, shown as triangles in the figure.

4. Conclusions

The results indicate that the dominant scattering mechanism will be intravalley scattering, with $|V_\Delta|^2$ being about a factor of 6 greater than intravalley alloy scattering in SiGe [5, 6]. In SiGe alloys, each scattering parameter is of comparable magnitude at around 0.5 eV. In the present case, f-type scattering is around 4 times weaker and g-type around 3 times stronger than in SiGe. Thus the application of strain, which would lift the degeneracy of perpendicular valleys, will have little effect on the carrier mobility. It is interesting to compare these present results to the much higher scattering due to substitutional nitrogen in GaInAs alloys [14]. Both C and N have much smaller covalent radii and much larger electronegativities than the atoms they substitute for, so in both case may be thought of as forming 'extreme alloys'. However, the results presented here suggest that extremely high carrier scattering, as observed in GaAs:N, is not necessarily a signature of such alloys.

Acknowledgments

The authors would like to thank Philip Murphy-Armando for many useful discussions. This work was supported by the Science Foundation Ireland.

References

[1] Jain S, Osten H, Dietrich B and Rücker H 1995 *Semicond. Sci. Tech.* **10** 1289–1302
[2] Osten H, Rücker H, Liu J and Heinemann B 2001 *Microelectron. Eng.* **56** 209–212
[3] Tan C, Chor E, Liu J, Lee H, Quek E and Chan L 2003 *Appl. Phys. Lett.* **83** 4169
[4] Osten H and Gaworzewski P 1997 *J. Appl. Phys.* **82** 4977–4981
[5] Murphy-Armando F and Fahy S 2006 *Phys. Rev. Lett.* **97** 96606
[6] Murphy-Armando F and Fahy S 2008 *Phys. Rev. B* **78** 35202
[7] Joyce S, Murphy-Armando F and Fahy S 2007 *Phys. Rev. B* **75** 155201
[8] Gonze X, Beuken J, Caracas R, Detraux F, Fuchs M, Rignanese G, Sindic L, Verstraete M, Zerah G, Jollet F *et al.* 2002 *Computational Materials Science* **25** 478–492
[9] Troullier N and Martins J 1991 *Physical Review B* **43** 1993–2006
[10] Broyden C 1967 *Math. Comput.* **21** 368–381
[11] Goldfarb D 1970 *Math. Comput.* **24** 23–26
[12] Shanno D 1970 *Mathematics of Computation* **24** 647–656
[13] Schlegel H 1982 *J. Comput. Chem.* **3** 214–218
[14] Skierbiszewski C, Perlin P, Wisniewski P, Suski T, Walukiewicz W, Shan W, Ager J, Haller E, Geisz J, Friedman D *et al.* 1999 *physica status solidi(b)* **216** 135–139

Journal of Physics: Conference Series **242** (2010) 012004 doi:10.1088/1742-6596/242/1/012004

Calculating Charged Defects using CRYSTAL

Christine L. Bailey[1], Leandro Liborio[2], Giuseppe Mallia[2], Stanko Tomić[1] and Nicholas M. Harrison[1,2]

[1]CSED, STFC Daresbury Laboratory, Cheshire, WA4 4AD, United Kingdom
[2]Department of Chemistry, Imperial College London, SW7 2AZ, United Kingdom

E-mail: christine.bailey@stfc.ac.uk

Abstract. The methodology for the calculation of charged defects using the CRYSTAL program is discussed. Two example calculations are used to illustrate the methodology: He^+ ions in a vacuum and two intrinsic charged defects, Cu vacancies and Ga substitution for Cu, in the chalcopyrite $CuGaS_2$.

1. Introduction

In recent years it has become possible to accurately calculate defect formation energies in crystalline solids using *ab initio* methodologies. While the calculation of uncharged defects is relatively straight-forward, the calculation of charged defects is more complicated [1]. In this paper we outline a methodology for the calculation of charged defect formation energies using the CRYSTAL program. The techniques used are transferable to other *ab initio* programs.

The first example is a very simple system, consisting of a periodic array of He^+ ions in a vacuum. The second example involves charged defects in the chalcopyrite $CuGaS_2$. The defects considered are the vacancy formed by the removal of a Cu^+ ion (V_{Cu}^-) and the substitution of a Cu^+ ion for a Ga^{3+} ion (Ga_{Cu}^{2+}). This example has been chosen as it is of current scientific interest, as $CuGaS_2$ could potentially be used as an absorber layer in highly efficient solar cells.

2. Calculational Scheme

All calculations are performed using the CRYSTAL program. This program computes the electronic structure of systems within Hartree-Fock, density functional (DFT) and various hybrid approximations using local Gaussian basis sets for systems periodic in three, two, one and no dimensions.

Our first example, a periodic array of He^+ ions in a vacuum, is calculated using Hartree-Fock theory and a standard Pople 3-21G basis set. The second example, V_{Cu}^- and Ga_{Cu}^{2+} defects in $CuGaS_2$, is of current scientific interest, consequently, care has been taken to perform calculations that will lead to accurate defect formation energies. A full account of this work can be found elsewhere [2]. The B3LYP hybrid exchange functional [3] is used. This functional has been shown to provide a reliable description of geometric and electronic structure and energetics in a wide range of materials [4, 5]. In particular, hybrid functionals, such as B3LYP, provide a much better prediction of the band gap of semiconductors than local density approximation or generalised gradient approximation DFT functionals. Polarised triple valence Gaussian basis sets were used.

© 2010 IOP Publishing Ltd

Journal of Physics: Conference Series **242** (2010) 012004 doi:10.1088/1742-6596/242/1/012004

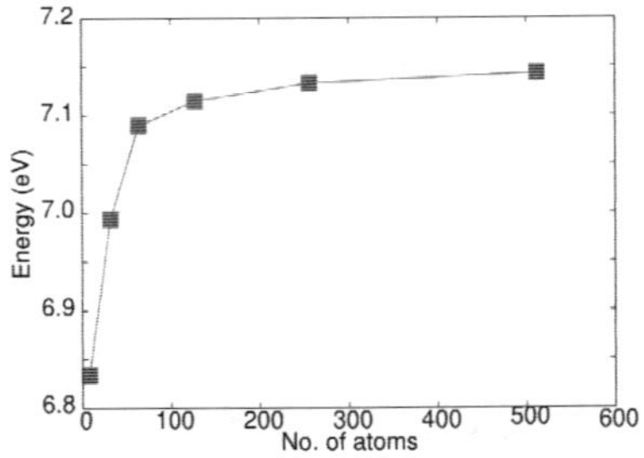

Figure 1. The change in the energy of the uncharged $CuGaS_2$ system when an electron is removed

3. Defect Formation Energies

The formation energy of a defect, D, of net charge q, in system X is defined as

$$\Delta H_f(D) = E_{\rm tot}(D) - E_{\rm tot}(X) + \sum_i n_i\mu_i + q(E_F - E_V) \tag{1}$$

where $E_{\rm tot}(D)$ and $E_{\rm tot}(X)$ are the total energies of the system with and without the defect, D. n_i represents the number of atoms of element i that are removed from the system when the defect is formed (a negative value for n_i denotes addition of atoms). μ_i is the chemical potential of element i, it represents the energy of the atoms that are removed (or added) to the system when the defect is formed. The fourth term, $q(E_F - E_V)$, represents the energy change due to the exchange of electrons and holes with the carrier reservoirs. $(E_F - E_V)$ is the Fermi energy relative to the valence band maximum (VBM) of the defect free system.

4. Energy Offset Correction

A consequence of using periodic boundary conditions within electronic structure calculations is that the boundary conditions lead to the conditional convergence of the Coulomb potential. In the case of uncharged systems the potential and total energy converge to well defined values under the conditions described first by Ewald [6]. The total energy of a charged system, however, can only be calculated to within a constant offset [1]. The value of this constant offset depends on the average crystal potential. It can be obtained by calculating the change in the energy of an uncharged system when an electron is removed from it; as the system size is increased the change in energy converges to the value of the offset.

In our first example, He^+ in a vacuum, the energy offset is equal to zero as the defect free system is simply a vacuum and the total energy of a uniform background charge in the limit of the charge tending to zero (equivalent to the system size tending to infinity) is zero. The change in energy of the $CuGaS_2$ system before and after the removal of an electron is displayed in figure 1 for increasing supercell size. As the system size tends towards infinity the difference in energy converges to approximately 7.15 eV. The value of $E_{\rm tot}(D)$ in equation 1 must be corrected by this constant offset multiplied by the net charge of the system.

5. Coulomb Interactions between Defects

The total energy of a periodic system that contains a localised charged defect treated within a finite supercell includes a term due to defect-defect, defect-background and background-background Coulomb interactions. To calculate the energy of an isolated defect this term must be subtracted from $E_{\rm tot}(D)$. It can be approximated by the multi-pole correction [7]:

$$\Delta E = \frac{q^2\alpha_M}{2\epsilon_r V^{\frac{1}{3}}} + \frac{2\pi q Q}{3\epsilon_r V} + O(V^{-\frac{5}{3}}), \tag{2}$$

where α_M is the lattice dependant Madelung constant, ϵ_r is the relative dielectric constant, V is the volume of the cell and Q is the quadrupole moment of the defect.

The first term in equation 2 represents the defect-defect Coulomb interaction. It can trivially be calculated using CRYSTAL as it is equivalent to the nuclear-nuclear interaction of a periodic system consisting of hydrogen atoms at the positions of the defects multiplied by q^2/ϵ_r. This nuclear-nuclear interaction energy is reported in the standard output of a CRYSTAL calculation. The value of ϵ_r in equation 2 can either be obtained from experimental results or calculated directly using CRYSTAL [8, 9]. The second term in equation 2 is due to the interaction between the defects and the background charge. The analytical calculation of the second term in equation 2 is not straight forward, hence it is often easier to obtain it numerically. In many cases, however, this term will be small enough to be neglected. The third term scales as $1/V^{5/3}$ and can almost always be neglected. In sections 5.1 and 5.2 the effect of including the first and second order terms in the calculation of defect energies are investigated.

5.1. Example 1: He$^+$ in a vacuum

The energy of an He$^+$ ion in a periodic cubic array was calculated for increasing cell size. After accounting for the electrostatic interactions described by equation 2, the energy of the periodic system should equal the energy of the isolated ion. For comparison, the energy of an isolated He$^+$ was calculated using CRYSTAL with no periodic boundary conditions. The difference in energy between the isolated He$^+$ ion and the periodic system, before and after accounting for the first order electrostatic interactions, is plotted in figure 2a, second order corrections are also included in figure 2b. A cubic cell of length 2Å leads to a total energy per He$^+$ ion that is more than 10 eV smaller than the energy of an isolated He$^+$ ion if the electrostatic corrections are not taken into account. After subtracting the energy given by the first term in equation 2 the difference in energy between the two systems is reduced to around 1 eV, applying second order corrections reduces the difference in energy to 0.02 eV.

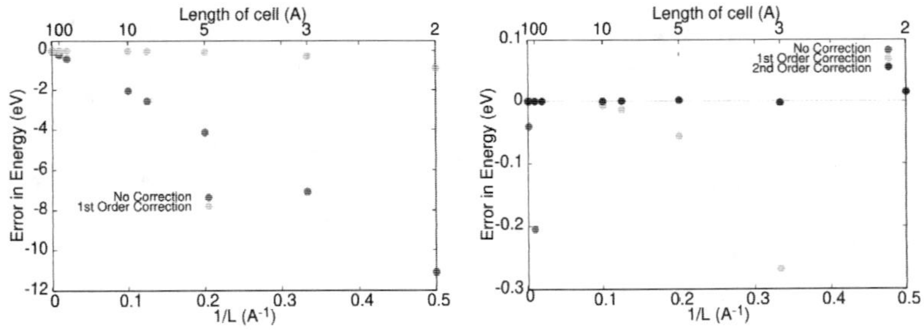

Figure 2. The difference in the calculated energy of an isolated He$^+$ ion and a periodic system consisting of an He$^+$ ion in each cell.

5.2. Example 2: V_{Cu}^- and Ga_{Cu}^{2+} defects in $CuGaS_2$

The formation energy of V_{Cu}^- and Ga_{Cu}^{2+} defects as a function of increasing supercell sizes has been calculated. To isolate the purely electrostatic effects these calculations were performed without geometry optimisation. The resultant energies before and after applying first order corrections (the first term in equation 2) are shown in Fig. 3. The straight lines in this figure represent the calculated energies for the largest supercell sizes. The addition of first order terms has led to a better convergence of the formation energies of the Ga_{Cu}^{2+} defect than the V_{Cu}^- defect. Analysis of the results reveals that the convergence of the V_{Cu}^- defect energy can be improved by multiplying the pre-factor of the $1/V^{1/3}$ correction term by 1.8. It is likely that this is because additional factors, such as elastic effects, that also affect the convergence of the formation energy [10, 11]. Geometry relaxation of the defects may also lead to changes in the convergence of the defect energies with respect to supercell size. The convergence of the Ga_{Cu}^{2+} and V_{Cu}^- defect formation energies can not be significantly improved through the addition of a $1/V$ correction term. It is notable that the formation energies calculated for the 8 atom supercell

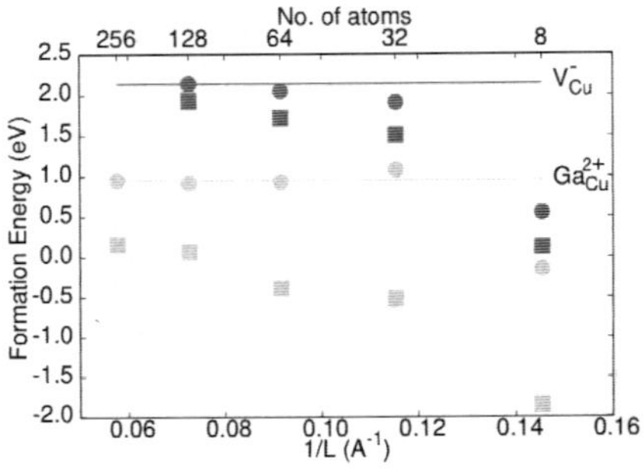

Figure 3. The formation energies of V_{Cu}^{-} (red) and Ga_{Cu}^{2+} (green) in $CuGaS_2$. The squares denote energies before any corrections are applied and the circles denote energies after first order corrections are applied.

are very poorly estimated, even after first order corrections. It is likely that this is because the defect charge is not fully localised within the supercell.

Ideally, the rate of convergence of a defect formation energy in a fully relaxed system should be analysed for increasing supercell sizes. In most cases, however, it will be computationally expensive to perform such tests. Analysis of our results suggests that the formation energy of a V_{Cu}^{-} defect in $CuGaS_2$ can be calculated to within an accuracy of 0.2 eV for a 64 atom supercell after including an energy offset correction (section 4) and first order corrections given by equation 2. The formation energy of a Ga_{Cu}^{2+} defect is calculated to an accuracy of around 0.05 eV. Previous studies that have considered a wide range of charged defects show that, after applying first order corrections, a 64 atoms supercell is usually sufficient to obtain charged defect formation energies to within 0.2 to 0.3 eV, and in many cases to within less than 0.1 eV [10].

6. Conclusion

The methodology for the calculation of charged defects has been discussed and illustrated through the use of two example calculations. It is shown that an energy offset correction must be applied to allow a comparison between the energies of charged and uncharged systems. Ideally the convergence of the formation energy of a charged defect should be analysed for increasing system size. It is, however, usually sufficient to perform a single energy calculation on a cell of approximate size $10\times10\times10$ Å^3 and estimate the energy term due to Coulomb interactions between periodically repeating charged defects to obtain a formation energy of an isolated defect to within about 0.1 eV.

References

[1] de Walle C G V and Neugebauer J 2004 *J. Appl. Phys.* **95** 3851
[2] Bailey C L, Liborio L, Mallia G, Tomić S and Harrison N M 2010 *Phys. Rev. B* **81** 205214
[3] Becke A D 1993 *J. Chem. Phys.* **98** 1372
[4] Muscat J, Wander A and Harrison N M 2001 *Chem. Phys. Letts.* **342** 397
[5] Tomić S, Montanari B and Harrison N M 2008 *Physica E* **40** 2125
[6] Ewald P 1921 *Ann. Phys.* **369** 253
[7] Makov G and Payne M C 1995 *Phys. Rev. B* **51** 4014
[8] Dovesi R, Saunders V R, Roetti C, Orlando R, Zicovich-Wilson C M, Pascale F, Civalleri B, Doll K, Harrison N M, Bush I J, D'Arco P and Llunell M 2007 *CRYSTAL 2006 Users's Manual* (University of Torino)
[9] Darrigan C, Rérat M, Mallia G and Dovesi R 2003 *J. Comp. Chem.* **24** 1305
[10] Castleton C W M, Hoglund A and Mirbt S 2009 *Modelling Simul. Mater. Sci. Eng.* **17** 084003
[11] Freysoldt C, Neugebauer J and de Walle C G V 2009 *Phys. Rev. Lett.* **102** 016402

2nd Workshop on Theory, Modelling and Computational Methods for Semiconductors IOP Publishing
Journal of Physics: Conference Series **242** (2010) 012005 doi:10.1088/1742-6596/242/1/012005

DFTB$^+$ and lanthanides

B Hourahine[1], B Aradi[2] and T Frauenheim[2]

[1] Department of Physics, SUPA, University of Strathclyde, John Anderson Building, 107 Rottenrow, Glasgow G4 0NG, United Kingdom
[2] BCCMS, Universität Bremen, Am Fallturm 1, 28359 Bremen, Germany

E-mail: benjamin.hourahine@strath.ac.uk

Abstract. DFTB$^+$ is a recent general purpose implementation of density-functional based tight binding. One of the early motivators to develop this code was to investigate lanthanide impurities in nitride semiconductors, leading to a series of successful studies into structure and electrical properties of these systems. Here we describe our general framework to treat the physical effects needed for these problematic impurities within a tight-binding formalism, additionally discussing forces and stresses in DFTB. We also present an approach to evaluate the general case of Slater-Koster transforms and all of their derivatives in Cartesian coordinates. These developments are illustrated by simulating isolated Gd impurities in GaN.

1. Introduction

The Kohn-Sham density functional theory (DFT) can be approximated to give a family of semi-empirical methods with much lower computational costs than full DFT. The Density Functional based Tight binding method (DFTB) [1] is a non-orthogonal method derived from an approximate second-order expansion of DFT with respect to fluctuations in the density [2]. This gives a transparent and parameter-free technique where all terms can be readily pre-calculated using DFT. This method has been applied with some success to a range of solid state, molecular and biological systems.

The study of systems like rare-earth or transition metal containing compounds and solids requires somewhat different techniques beyond charge self-consistent DFTB [3]. Spin polarisation [4] is essential, and for the strongly correlated d and f electrons, methods like LDA+U [5] or pseudo-SIC [6] also become necessary. Additionally lanthanide atomic numbers are between 57 and 70, hence relativistic effects start becoming important. The latest version of the DFTB$^+$ code [7] includes extensions to address all of this additional complexity in the treatment of lanthanides, the DFTB method having been substantially extended by adapting LDA+U and pseudo-SIC into a common framework [8] and adopting Russel-Saunders $L \cdot S$ coupling [4]. Here we describe the approach used to include these disparate effects within an common computational framework.

1.1. non-SCC DFTB and Slater-Koster transforms

In the present case, Slater-Koster type tight binding [9] requires geometrical transformations for two centre terms between orbitals with angular momentum of up to f. Closed form expressions for these transforms to arbitrary angular momentum [10] and their analytical derivatives [11] are available. However, in the spherical-polar coordinate systems used, there are spurious

© 2010 IOP Publishing Ltd

singularities at the poles, leading to these points being presented as separate special cases without a continuous functional form over the whole coordinate range. The transformations are written in terms of directional cosines between atomic pairs (N, M and L) and depend on products of the general form $exp(i\phi|m|)d^l_{m,m'}(\theta)$ where l, m and m' are the usual angular labels, θ and ϕ are the polar and azimuthal angles (see Ref. [10] for details). The apparently problematic part of the Wigner-d matrices in these terms is

$$d^l_{m,m'} \propto \sum_{t=0,non-neg}^{2\times l+1} (1-N)^{t+m/2+m'/2}(1+N)^{l-t-m/2+m'/2} \times \frac{(-1)^t}{(l+m'-t)!(l-m-t)!t!(t+m-m')!}$$

where the summation over t is such that the factorial terms in the denominator are all $\geqslant 0$. This summation constraint leads to the $(1 \pm N)$ terms being well-behaved even at the poles ($|N|=1$). This holds from $t \geqslant 0$ and $t \geqslant m'-m$, $\therefore t+m/2+m'/2 \geqslant 0$ and similarly, $l+m'-t \geqslant 0$ and $l-m-t \geqslant 0$, $\therefore l-m/2+m'/2-t \geqslant 0$.

However the problem of the azimuthal angle, which becomes undefined at either pole, remains. This can be partially alleviated by defining this angle using $\phi = atan2(M,-L)$, but care is required since $atan2(0,0)$ is again undefined (but in specific machine implementations, Intel's FPATAN [12] for example, it is defined as the required case of ± 0 for sign choices of $atan2(\pm 0,+0)$). The overall exponential term is then of the form $\left((M+iL)/\sqrt{M^2+L^2+\delta}\right)^{|m|}$ where delta is infinitesimally positive, and again the $M=L=0$ case requires a convention that the function gives +1.

Since the system as a whole is invariant under the rotation group O_3, a simpler option is to perform the calculation in a global coordinate system rotated, for example, in the xz plane and chosen such that no inter-atomic vectors align exactly along the z axis (in all cases $|N| \neq 0$). This allows the original rotation expressions and any order of their derivatives to be used. In the situation that a change in geometry brings an atomic pair into exact alignment with $\pm z$, the stable Wigner d-matrix form given above can then be used to rotate any orientationally dependent terms in the Hamiltonian away from the problematic alignment by an additional application of Eqns. 5-6 of Ref [10] with the azimuthal change, γ, chosen as 0. However, in practice, DFTB$^+$ currently uses explicitly coded cases of the geometrical transformations up to f.

1.2. Potential shifts

The total energy for spin-polarised SCC DFTB is the following functional of the density matrix ρ:

$$E = \sum_{\mu\nu} H^0_{\mu\nu}\rho_{\nu\mu} + \frac{1}{2}\sum_{ab}\left(\delta q_a[\rho]\gamma_{ab}\delta q_b[\rho] + \delta m_a[\rho] W_{ab}\delta m_b[\rho]\right)$$

where δq and δm are fluctuations in the charge and magnetisation respectively from the reference system used to construct the non-self-consistent Hamiltonian, H^0. The domains a and b are Mulliken populations constructed from the density and overlap (S) matrices, $q_\mu = \frac{1}{2}\sum_\nu S_{\mu\nu}\rho_{\nu\mu}$, where q_μ is additionally summed over either whole atoms or atomic shells with a given value of l to give q_a. The couplings γ and W are derived from either an approximation to the Hartree and exchange correlation or an on-site Stoner-like exchange contribution respectively between the fluctuations. The reference system chosen is a superposition of neutral and spin-free atomic densities (or potentials, see Ref. [2] for more details of conventional DFTB).

The Hamiltonian for the system, derived using the usual approach of minimising the total energy subject to orthonormality of the resulting single particle states, is then of the general form

$$H^\sigma_{\mu\nu} = H^0_{\mu\nu} + \frac{1}{2}S_{\mu\nu}\left(\Omega^\sigma_\mu + \Omega^\sigma_\nu\right)$$

where the spin dependent shifts are defined as $\Omega^\sigma_\mu = \sum_b \gamma_{\mu b}\delta q_b[\rho] + \sigma \sum_b W_{\mu b}\delta m_b[\rho]$ for spins $\sigma = \pm 1$. More general Hamiltonian contributions can also be cast into this form. An example would be the constrained density method of Dederichs *et al.* [13] where the ground state density matrix is found for a chosen charge distribution (so similar to the Levy–Lieb constrained search method), by finding the ground state of the system in a (carefully chosen) external potential. Wu and Van Voorhis [14] formulated this as a maximisation problem for integral constraints using undetermined multipliers, by defining a new functional:

$$W[n(\mathbf{r}), \mathbf{V}] = E[n(\mathbf{r})] + \sum_i V_i \left(\int w_i(\mathbf{r}) n(\mathbf{r}) dr - N_i \right)$$

to constrain N_i electrons to match the weighting envelope $w_i(\mathbf{r})$ over the density or magnetisation within region that the i-th constraint is formulated. W is convex in $n(\mathbf{r})$ and concave in $\mathbf{V}$. This then also gives a contribution to the DFT potential. To write these constraints for DFTB's Mulliken-charge based tight-binding populations is straight forward, and again gives a stationary energy structure, as with DFT.

$$W[\rho, \mathbf{V}] = E_{\mathrm{DFTB}}[\rho] + \sum_i V_i \left(\sum_\mu w_{i\mu} q_\mu[\rho] - N_i \right)$$

where μ are basis function labels for the ith constraint, giving an equivalent shift contribution of the form

$$\Omega_\mu = \sum_i V_i \sum_\nu w_{i\nu} S_{\mu\nu}$$

More general constraints of the form $U_i \left(w_{i\mu}q_\mu[\rho] - N_i\right)^\kappa$ can also be used, where in the case that κ is odd the constraint is enforced by the choice of U as a Lagrange undetermined multiplier, while for even values of κ this constraint is approximately enforced as $U \to \infty$.

1.3. Generalisations for orbital potentials and $L \cdot S$

Instead of choosing Mulliken populations as our basic variable, the more general on-site occupation matrix [15] can be used for assorted semi-local non-mean field potentials. These matrices are defined as

$$n_{\mu\nu} = \frac{1}{2}\sum_\tau \left(S_{\mu\tau}\rho_{\tau\nu} + \rho_{\mu\tau}S_{\tau\nu}\right); \ \mu\nu \in a$$

where the matrix diagonal is the conventional Mulliken charge for atom a. Further generalising, the Hamiltonian and occupation matrices can be projected onto the orthogonal Pauli matrices, giving charge and non-collinear vectoral magnetisation in the system. The spin-block Hamiltonian and density matrix can then be written as coefficients of Pauli matrices ($\boldsymbol{\sigma}_{I,x,y,z}$):

$$H_{\mu\nu} = H^0_{\mu\nu}\boldsymbol{\sigma}_I + \frac{1}{2}S_{\mu\nu} \sum_{j=x,y,z} \left(\Omega^j_\mu + \Omega^j_\nu\right)\boldsymbol{\sigma}_j$$

$$\rho = \sum_{j=I,x,y,z} \rho^j_{\nu\mu}\boldsymbol{\sigma}_j$$

Journal of Physics: Conference Series **242** (2010) 012005 doi:10.1088/1742-6596/242/1/012005

for the corresponding generalised vectoral shifts. In the case of Russel-Saunders spin-orbit coupling (or Complex Absorbing Potentials [16]) these Pauli-matrix coefficients become complex.

1.4. Forces and stresses

First derivatives of the band-structure part of the total energy can similarly be written in terms of the shift components, using the Hellman-Feynman theorem and the Pulay correction for the change of basis (using the energy weighted density matrix, ρ^E). The ith component of the derivative of energy with respect to the position of atom a due to atom b is

$$\left.\frac{\partial E}{\partial x_{ai}}\right|_b = \sum_{\mu\in a}\sum_{\nu\in b}\frac{\partial H^0_{\mu\nu}}{\partial x_{ai}}\rho_{\nu\mu} + \sum_{\mu\in a}\sum_{\nu\in b}\frac{\partial S_{\mu\nu}}{\partial x_{ai}}\left(\left(\Omega_\mu + \Omega_\nu\right)\rho_{\nu\mu} - \rho^E_{\nu\mu}\right)$$

where the ith Cartesian component of the force on atom a is then

$$F_{ai} = -\sum_b \left.\frac{\partial E}{\partial x_{ai}}\right|_b$$

and the summation is calculated using the real-space matrices of Ref. [7]. The stress tensor is also of similar form, with band-structure contributions of

$$\sigma_{ij} = -\frac{1}{2V_{\text{cell}}}\sum_{ab}\left.\frac{\partial E}{\partial x_{ai}}\right|_b (x_{aj} - x_{bj})$$

where V_{cell} is the unit cell volume. For the total stress and forces, the repulsive energy and also long range interactions such as the electrostatic double counting and the dispersive interactions introduce other terms into the total expressions, but these are of standard forms from the literature.

The two main advantages of using the shift contributions formulation of the forces are that firstly it simplifies the implementation of the majority of the force terms (since the code is then common for many distinct contributions) using the already calculated shift contributions from the Hamiltonian. Secondly it provides a semi-independent test for the correctness of any new extension to DFTB, since while the Hamiltonian and the band-structure part of the forces depend on the shifts, it is generally not possible to calculate the total energy from the total shift, due to the differing functional forms of the various contributions. Therefore agreement between the numerical and analytical derivatives of the energy signify that the independent shift and total energy terms agree (and therefore the shift is the derivative of the energy with respect to the eigenstates of the system).

2. Gd doped GaN

Focussing on a specific application of this formalism we now consider treatment a specific substitutional rare-earth impurity in gallium nitride. RE_{Ga} substitutionals are the simplest stable lanthanide defects in GaN [17, 18, 19, 20, 21, 22, 23]. From experimental studies we know that the ions prefer the Ga position [24], occur in the 3+ valence state [25] and possess C_{3v} symmetry [26] with relatively short distances to the surrounding N-ligands [27]. Gadolinium doped GaN has been suggested to [28] have an extraordinarily large magnetic moment per dopant atom.

The local structure of the substitutional is insensitive as to whether the Ga$-3d$ orbitals are treated as valence or core. The results here use a parameterization where the Ga-$3d$ states are not included in the valence, but these do not differ substantially from results obtained by including them [17]. Supercells containing 192 atoms and a 6^3 Monkhorst-Pack k-point sampling [29] were used to calculate the results reported here, with the density of states for the bulk GaN supercell used shown in Fig. 1, with the density inverted in sign above the Fermi energy (this convention is also used in other plots where two spin channels are shown). The substitutional Gd_{Ga} in the lowest energy configuration is found to have C_{3v} symmetry as expected with Gd–N distances of 2.246 and 2.242 Å for the axial and 3 fold surrounding bonds respectively.

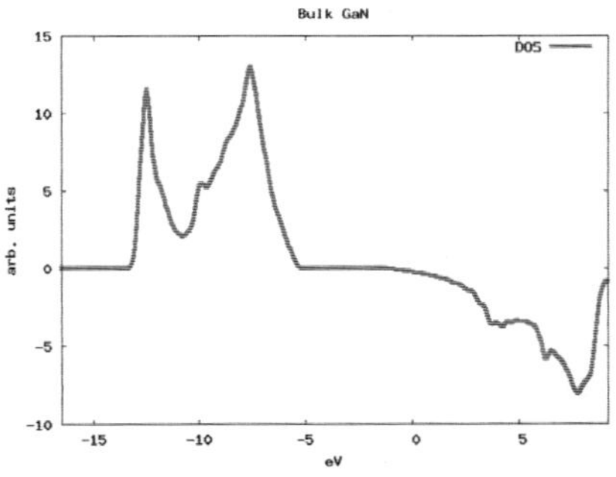

Figure 1. Bulk DOS of GaN, empty states are marked below 0.

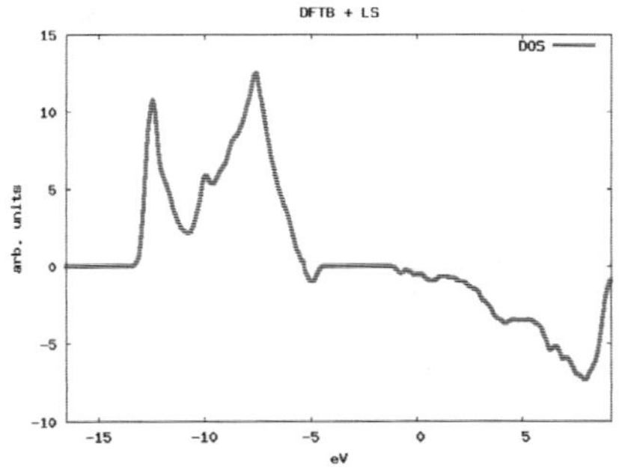

Figure 2. Non-collinear z-aligned density of states for Gd doped GaN including spin-orbit coupling.

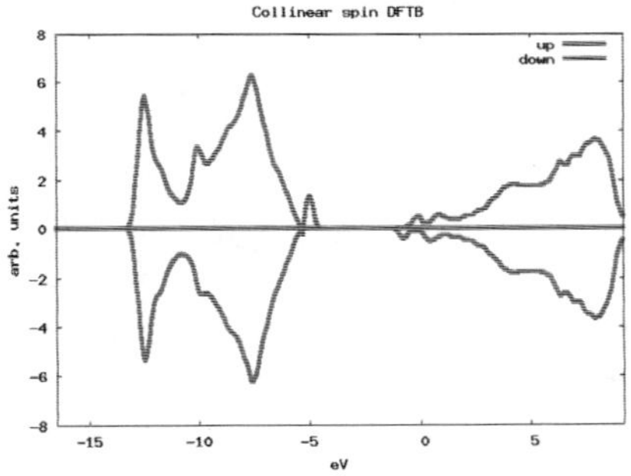

Figure 3. Collinear density of states for Gd doped GaN.

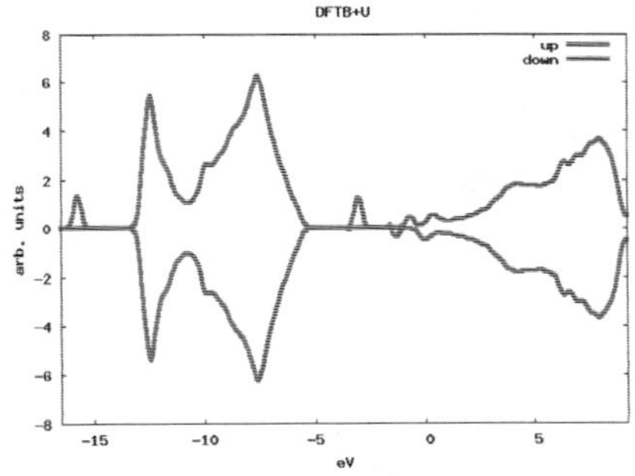

Figure 4. DFTB+U collinear density of states for Gd doped GaN.

The conventional spin polarised DFTB calculation leads to 7 unpaired spins primarily localised on the Gd (6.77) with 96 % of that localized on the $4f$ shell. As shown in figure 3 the presence of the Gd ions splits the spin degeneracy of the bottom of the conduction band and would provide ferromagnetic coupling in the case of strong n-type doping [30], however this does not explain the observed moment in the experimental conditions of highly resistive material [28].

The collinear (z) spin-polarized and non-collinear calculations are degenerate in energy. The system is, as expected, degenerate with respect to rotation of the spin direction in the non-collinear case. The addition of spin orbit coupling splits this degeneracy (values of the screened atomic spin-orbit constants, ξ, are taken from Refs. [31, 32]), but by less than 1 meV. The $4f$ shell is strongly spin polarized with 6.5 μ_B unpaired spin and a parallel orbital momentum of 0.28 μ_B, yielding a total moment of 6.8 μ_B, all similar to the expected $^8S_{7/2}$ ground state. As shown in Fig. 2 this does not lead to a substantial change in the density of states compared to the collinear spin polarised case.

The $4f$ states of Gd are present around the valence band maximum for the spin-polarised calculations, due to the incorrect treatment of the self-interaction of these states. Applying LDA+U type potentials [18] shifts this type feature, moving the occupied states downwards in energy, and depending on the potential, the unoccupied states upwards.

Previous application of DFTB+U to other rare-earths and their complexes [33] have demonstrated the applicability of this method to these systems. In the specific case of Gd_{Ga} the "FLL" LDA+U functional has been applied with a choice of 7.43 eV for $(U - J)$ in accordance with Ref. [8]. This leads to the an increase in the total energy by $\sim$0.8 eV, consisting primarily of an increase of 1.4 and 0.3 eV in the band structure and additional LDA+U energy terms, but a reduction of 0.9 eV in the spin polarization energy, due primarily to the increased localization of the unpaired spin on the Gd $4f$ shell. This is evident between Figs. 3 and 4 where the empty $4f$ states are lifted clear from the valence band, while the filled states drop below the region of the valence band.

3. Acknowledgements

We thank Simone Sanna for making available his unpublished parameterization of Gd doped GaN. This work was partly funded by the FP7 ITN "RAINBOW" (PITN-GA-2008-213238)

References

[1] Porezag D, Frauenheim T, Köhler T, Seifert G and Kaschner R 1995 *Phys. Rev. B* **51** 12947
[2] Frauenheim T, Seifert G, Elstner M, Niehaus T, Kohler C, Amkreutz M, Sternberg M, Hajnal Z, Di Carlo A and Suhai S 2002 *J. Phys. Cond. Matter* **14** 3015–3047
[3] Elstner M, Porezag D, Jungnickel G, Elsner J, Haugk M, Frauenheim T, Suhai S and Seifert G 1998 *Phys. Rev. B* **58** 7260–7268
[4] Kohler C, Frauenheim T, Hourahine B, Seifert G and Sternberg M 2007 *J. Phys. Chem. A* **111** 5622–5629
[5] Anisimov V I, Aryasetiawan F and Lichtenstein A I 1997 *J. Phys. Cond. Matter* **9** 767
[6] Filippetti A and Spaldin N A 2003 *Phys. Rev. B* **67** 125109
[7] Aradi B, Hourahine B and Frauenheim T 2007 *J. Phys. Chem. A* **111** 5678–5684
[8] Hourahine B, Sanna S, Aradi B, Koehler C, Niehaus T and Frauenheim T 2007 *J. Phys. Chem. A* **111** 5671–5677
[9] Slater J C and Koster G F 1954 *Phys. Rev.* **94** 1498
[10] Podolskiy A V and Vogl P 2004 *Phys. Rev. B* **69** 233101
[11] Elena A M and Meister M 2005 *Phys. Rev. B* **72** 165107
[12] 2004 *IA-32 Intel Architecture Software Developer's Manual*
[13] Dederichs P H, Blügel S, Zeller R and Akai H 1984 *Phys. Rev. Lett.* **53** 2512–2515
[14] Wu Q and Van Voorhis T 2005 *Phys. Rev. A* **72** 024502
[15] Han M J, Ozaki T and Yu J 2006 *Phys. Rev. B* **73** 045110
[16] Manolopoulos D E 2002 *J. Chem. Phys.* **117** 9552–9559
[17] Sanna S, Hourahine B, Gallauner T and Frauenheim T 2007 *J. Phys. Chem. A* **111** 5665–5670
[18] Sanna S, Hourahine B, Gerstmann U and Frauenheim T 2007 *Phys. Rev. B* **76** 155128
[19] Filhol J, Jones R, Shaw M and Briddon P 2004 *Appl. Phys. Lett.* **84** 2841–2843
[20] Dorenbos P and van der Kolk E 2006 *Appl. Phys. Lett.* **89** 061122
[21] Svane A, Christensen N E, Petit L, Szotek Z and Temmerman W M 2006 *Phys. Rev. B* **74** 165204
[22] Sanna S, Hourahine B, Frauenheim T and Gerstmann U 2008 *Phys. Status Solidi C* **5** 2358–2360
[23] Hourahine B, Sanna S, Aradi B, Kohler C and Frauenheim T 2006 *Physica B* **376** 512–515
[24] Wahl U, Alves E, Lorenz K, Correia J, Monteiro T, De Vries B, Vantomme A and Vianden R 2003 *Mater. Sci. Eng. B* **105** 132–140
[25] Torvik J, Feuerstein R, Pankove J, Qiu C and Namavar F 1996 *Appl. Phys. Lett.* **69** 2098–2100
[26] Glukhanyuk V, Przybylinska H, Kozanecki A and Jantsch W 2004 *Phys. Status Solidi A* **201** 195–198
[27] Katchkanov V, Mosselmans J, O'Donnell K, Nogales E, Hernandez S, Martin R, Steckl A and Lee D 2006 *Opt. Mater.* **28** 785–789
[28] Dhar S, Brandt O, Ramsteiner M, Sapega V F and Ploog K H 2005 *Phys. Rev. Lett.* **94** 037205
[29] Monkhorst H J and Pack J D 1976 *Phys. Rev. B* **13** 5188
[30] Dalpian G M and Wei S H 2005 *Phys. Rev. B* **72** 115201
[31] Łusakowski A 2005 *Phys. Rev. B* **72** 094429
[32] Cardona M and Christensen N E 2000 *Solid State Commun.* **116** 421 – 425
[33] Sanna S, Schmidt W G, Frauenheim T and Gerstmann U 2009 *Phys. Rev. B* **80** 104120

Journal of Physics: Conference Series **242** (2010) 012006 doi:10.1088/1742-6596/242/1/012006

Model of the coherent magneto-optical dynamics in a single chiral carbon nanotube

Gabriela Slavcheva and Philippe Roussignol*

Blackett Laboratory, Imperial College London, Prince Consort Road, London SW7 2AZ, UK
*Laboratoire Pierre Aigrain, Ecole Normale Supérieure, 24 rue Lhomond,
75231 Paris Cedex 5, France

E-mail: `g.slavcheva@imperial.ac.uk`

Abstract. A theoretical framework and dynamical model for description of the natural optical activity and Faraday rotation in an individual chiral single-walled carbon nanotube in the highly nonlinear coherent regime is proposed. The model is based on a discrete-level representation of the optically active states near the band edge. Chirality is modelled by a system Hamiltonian corresponding to energy-level configurations, specific for each handedness, that are mirror reflections of each other. An axial magnetic field is introduced through the Aharonov-Bohm and Zeeman energy-level shifts.The time evolution of the quantum system is studied using the coherent vector Maxwell-pseudospin equations. Giant natural and magneto-optical gyrotropy, exceeding the one of the artificial photonic metamaterials, is numerically demonstrated for a single (5,4) carbon nanotube and an estimate of the magnitude of the natural circular dichroism and specific optical rotatory power is obtained. The model provides a framework for investigation of chirality and magnetic field dependence of the ultrafast nonlinear optical response of a single carbon nanotube.

1. Introduction

Chirality is one of the main symmetries of the carbon nanotube geometry which determines the optical properties of single-walled carbon nanotubes (SWCNTs). The interaction of polarised light with chiral materials in the absence of magnetic fields gives rise to the phenomenon of natural optical (rotation) activity, whereby the polarisation plane is rotated continuously during the light propagation across the medium.

Time-resolved magnetic circular dichroism and magneto-optical rotatory dispersion techniques offer spectroscopic information which is different or impossible to obtain by other means. A detailed understanding of the mechanisms underlying the optical and magneto-optical birefringence, circular dichroism and rotation in the nonlinear coherent regime require an adequate theoretical description. On the other hand, formulation of a theory and model of the optical activity in chiral molecules, such as individual single-wall carbon nanotubes (SWCNTs), in the high-intensity nonlinear coherent regime and under an axial magnetic field, is of special interest from a fundamental point of view. To our knowledge, no such theory has been proposed and very little is known about the polarisation dynamics of the nonlinear optical and magneto-optical response of a single carbon nanotube. Exploiting these ultrafast nonlinear effects would open up pathways for the development of a novel class of ultrafast polarisation-sensitive integrated optoelectronic devices, based on single carbon nanotubes.

© 2010 IOP Publishing Ltd

2. Theoretical background

Carbon nanotubes exist in two (AL) left- ($n>m$) and (AR) right-handed ($n<m$) helical forms (enantiomers) depending on the rotation of two of the three armchair (A) chains of carbon atoms counterclockwise or clockwise when looking against the nanotube z-axis. The electronic band structure [1] is described by the quantization of the wavevector along the tube circumference, resulting in subbands in the valence and conduction band labelled by the quasi-angular momentum quantum number (μ).

We model the single chiral nanotube band-edge structure at the K point of the Brillouin zone by an ensemble of identical four-level systems, corresponding to the dipole optically allowed transitions for AL and AR nanotube enantiomers (Fig. 1). Absorption of right circularly polarised light σ^+ of resonance frequency ω_0 excites the electronic transitions $\mu-1 \rightarrow \mu$ in AL-handed SWCNTs and $\mu \rightarrow \mu-1$ transitions in AR-handed SWCNTs, the absorption of left circularly polarised σ^- light excites $\mu \rightarrow \mu-1$ transitions in AL-handed and $\mu \rightarrow \mu+1$ transitions in AR-handed SWCNTs [2] (Fig. 1). When a circularly polarised (in the x-y plane)

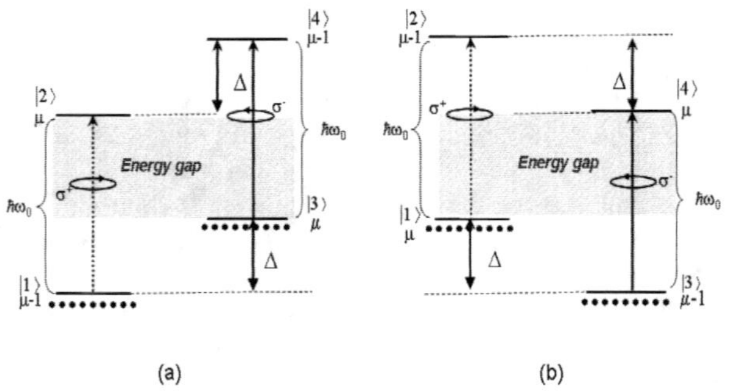

Figure 1. Energy-level structure at the $K(K')$ point of the lowest subbands labelled by the subband index μ for an (a) AL-handed; (b) AR-handed SWCNT. Δ is the energy separation between the lowest subband and the second lowest subband near the band edge.

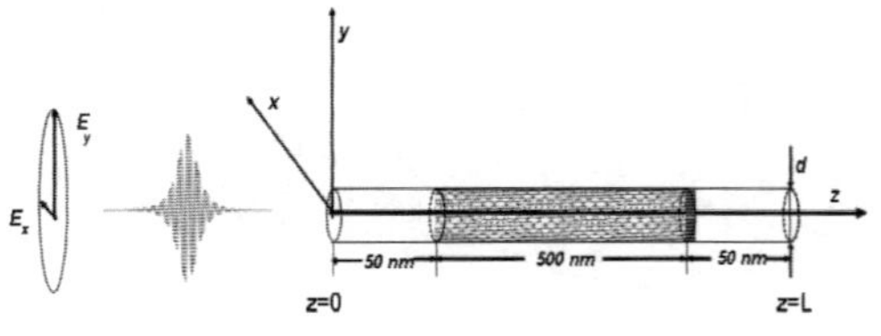

Figure 2. Geometry of an experiment with an optical excitation by circularly polarised pulse propagating along the SWCNT axis, E_x and E_y are E-field vector components.

laser pulse propagates along the z-axis of an AL or AR SWCNT (Fig. 2), only one of the two allowed transitions for a linearly polarised light (along x or y) is excited (see Fig.1).

2.1. Coherent dynamics of the natural optical activity

We follow the semiclassical formalism developed in [3] and describe the discrete multi-level system coherently driven dynamics by Maxwell's equations in vectorial form for the polarised optical pulse propagation coupled to master pseudospin equations for the time-evolution of the quantum system, exploiting the SU(4) group symmetries. We shall consider the specific case of a chiral AL-handed (5, 4) SWCNT, chosen for illustration of the general method valid for an arbitrary chirality. The electronic band structure, calculated by tight-binding method [4], can be well approximated by a four-level system, for which the system Hamiltonian is constructed

Journal of Physics: Conference Series **242** (2010) 012006 doi:10.1088/1742-6596/242/1/012006

according to [3]:

$$\hat{H} = \hbar \begin{pmatrix} 0 & -\frac{1}{2}\left(\Omega_x + i\Omega_y\right) & 0 & 0 \\ -\frac{1}{2}\left(\Omega_x - i\Omega_y\right) & \omega_0 & 0 & 0 \\ 0 & 0 & \Delta & -\frac{1}{2}\left(\Omega_x - i\Omega_y\right) \\ 0 & 0 & -\frac{1}{2}\left(\Omega_x + i\Omega_y\right) & \Delta + \omega_0 \end{pmatrix} \quad (1)$$

where $\Omega_x = \wp\frac{E_x}{\hbar}$, $\Omega_y = \wp\frac{E_y}{\hbar}$ are the time-dependent Rabi-frequencies and $\wp$ is the optical dipole matrix element. An estimate for the dipole transition matrix element in $(5,4)$ nanotube is obtained by using the tight-binding Hamiltonian, including chirality effects derived in [5], yielding a dipole moment $\wp = 3.613 \times 10^{-29}$ Cm.

The time evolution of the quantum system in the presence of relaxation processes is governed by a set of pseudospin master equations for the real state 15-dimensional vector S derived in [3]:

$$\frac{\partial S_j}{\partial t} = \begin{cases} f_{jkl}\gamma_k S_l + \frac{1}{2}Tr\left(\hat{\sigma}\,\hat{\lambda}_j\right) - \frac{1}{T_j}\left(S_j - S_{je}\right), & j = 1,2,...,12 \\ f_{jkl}\gamma_k S_l + \frac{1}{2}Tr\left(\hat{\sigma}\,\hat{\lambda}_j\right), & j = 13,14,15 \end{cases} \quad (2)$$

where γ is the torque vector, f is the fully antisymmetric tensor of the structure constants of SU(4) group, and T_j are the phenomenologically introduced non-uniform spin decoherence times describing the relaxation of the real state vector components S_j toward their equilibrium values S_{je}. The one-dimensional Maxwell's curl equations for the circularly polarised laser

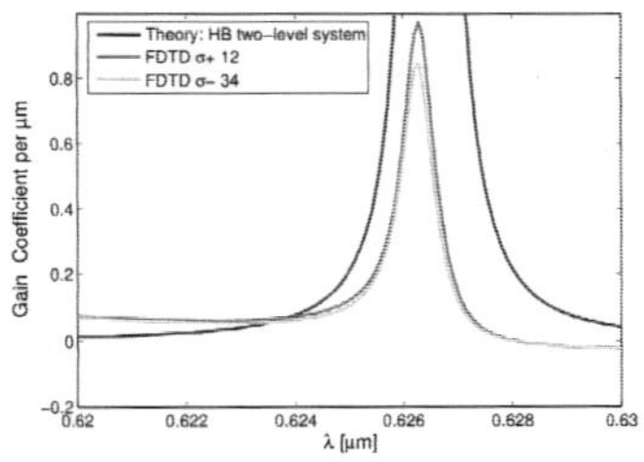

Figure 3. Calculated gain coefficient spectrum for σ^+ and σ^- ultrashort optical excitation

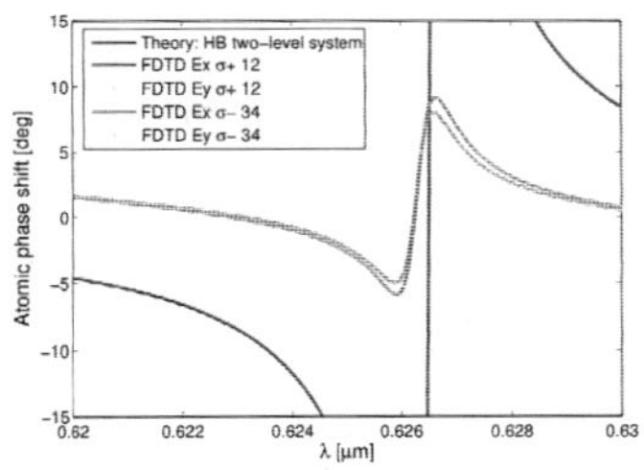

Figure 4. Calculated phase shift spectrum for σ^+ and σ^- ultrashort optical excitation

pulse propagating in an isotropic medium with an effective dielectric constant are solved self-consistently with master pseudospin equations Eq.(2). Employing the effective medium approximation [6], an effective dielectric constant of the $(5,4)$ nanotube is introduced, yielding a value of $\epsilon = 5.33$, or $n \approx 2.3$ for the refractive index. The initial circularly-polarized optical pulsed excitation, applied at the left boundary $z = 0$ of the simulation domain, shown in Fig. 2, is modelled by two orthogonal linearly polarized optical waves, phase-shifted by $\pi/2$.

The resulting set of equations is solved directly in the time domain, by employing the Finite-Difference Time-Domain method, yielding the spatially resolved temporal dynamics of the electric field vector components and the populations of all four levels along the nanotube length. The spatially resolved absorption/gain (Fig. 3) and phase shift spectra (Fig. 4) allow to calculate the circular dichroism and the optical rotation angle. The calculated average circular dichroism along the nanotube length is 0.083 μm^{-1}. An estimate of the circular dichroism of a (5,4) nanotube can be calculated from [5], giving a value of 1.03 for the relative difference between

σ^+ and σ^- absorption probability rates. This value, however, cannot be directly compared with the obtained result since the nanotube length does not enter the model in [5], and thus can only provide a guidance. We should stress that our analysis applies to the high-intensity nonlinear ultrafast regime, in which, up to our knowledge, no theory or experiments have been published. Therefore we should expect deviations from the linear case. By comparison, the absolute value of the circular dichroism of an artificial helicoidal bilayered structure varies in the range of 5-9 dB, which is equivalent to linear amplitude absorption/gain coefficient $0.58 - 1.04\ \mu\text{m}^{-1}$. The average natural specific rotatory power $\sim 2962.24°/\text{mm}$ exceeds the giant gyrotropy reported in the artificial photonic metamaterials of up to $2500°/\text{mm}$ [7]. We should note, however, that the complexity of the carbon nanotube molecular structure allows for engineering the optical activity in a wide range.

2.2. Magneto-optical activity of a chiral SWCNT in an axial magnetic field

In the presence of an axial magnetic field the electronic band structure of a single carbon nanotube, and the electronic states near the band gap edge in particular, significantly change, owing to the combined action of two effects: the spin-B interaction resulting in Zeeman splitting of the energy levels and the appearance of the Aharonov-Bohm phase in the wave function resulting in an additional energy level shift. The energy level configuration in an external magnetic field can be split into two reduced sets of levels, each of which represents a mirror image of the other, however the symmetry is broken by the allowed optical transitions in each case. We apply the formalism developed in 2.1 and calculate the spatially resolved optical absorption/gain and phase shift spectra, shown in Figs 5 and 6, respectively. It is worth noting that the magnetic

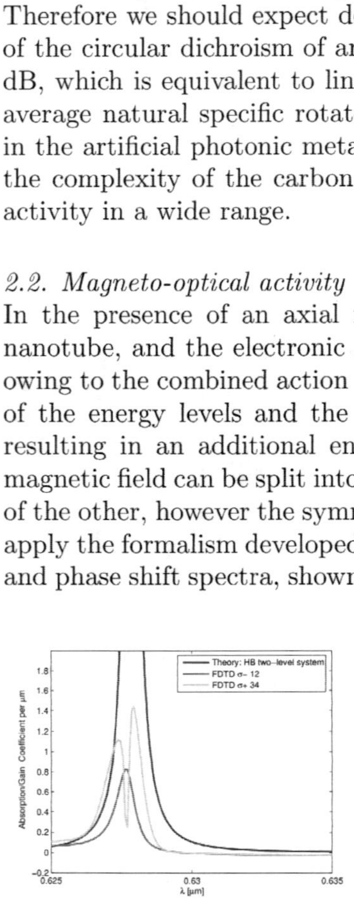

Figure 5. Calculated gain coefficient spectrum for σ^+ and σ^- ultrashort optical excitation at $B = 8$ T

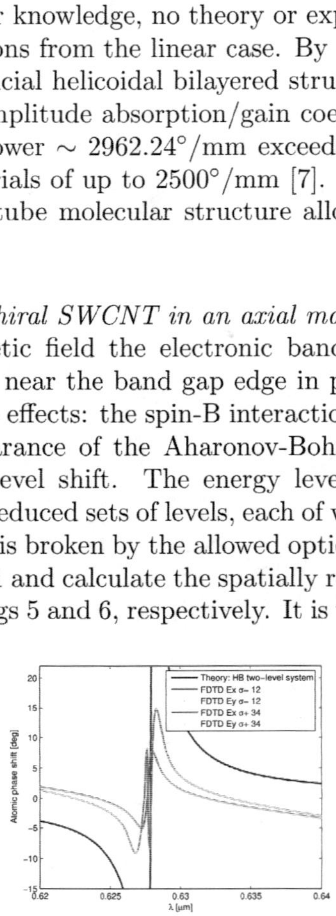

Figure 6. Calculated phase shift spectrum for σ^+ and σ^- ultrashort optical excitation at $B = 8$ T

circular dichroism spectra are quite distinctive from the zero-magnetic field ones. While the σ^- polarised pulse is amplified during the pulse propagation across the nanotube, the σ^+ polarised pulse is absorbed, resulting in a much larger net circular dichroism. The average value of the magnetic circular dichroism is $0.706\ \mu\text{m}^{-1}$, an order of magnitude larger than the natural circular dichroism. This is a direct consequence of the four-level energy-level system for a σ^- excitation and a three-level Λ-system for a σ^+ excitation. The calculated spectra in the latter case are reminiscent of electromagnetically-induced transparency (EIT) and coherent population trapping effects. The calculated average specific rotatory power in a magnetic field ($B = 8$ T), is $-32.5804°/\mu m$, the meaning of the minus sign is to denote left rotation. The predicted Faraday rotation at $B = 8$ T is nearly an order of magnitude greater than the natural optical rotation. We should note that the calculated rotation is a combined effect from the chirality of the nanotube and the magnetic field-induced rotation and thus can be considered as an estimate for the magneto-chiral effect in a single nanotube.

Journal of Physics: Conference Series **242** (2010) 012006 doi:10.1088/1742-6596/242/1/012006

References

[1] Ajiki H and Ando T 1993 *J. Phys. Soc. Jpn* **62** 1255
[2] Samsonidze Ge G *et al.* 2004 *Phys. Rev.* B **69** 205402
[3] Slavcheva G 2008 *Phys. Rev.* B **77** 115347
[4] Yang L, Anantram M P, Han J, and Lu J P 1999 *Phys. Rev.* B **60** 13874
[5] Ivchenko E L, and Spivak B 2002 *Phys. Rev.* B **66** 155404
[6] Henrard L and Lambin Ph 1996 *J. Phys.* B: At. Mol. Opt. Phys. **29** 5127
[7] Plum E, Fedorov V A, Schwanecke A S, Zheludev N I, and Chen Y 2007 *Appl. Phys. Lett.* **90** 223113

Journal of Physics: Conference Series **242** (2010) 012007 doi:10.1088/1742-6596/242/1/012007

Calculation of the few-electron states in semiconductor carbon nanotube quantum dots by exact diagonalisation

Mervyn Roy and P A Maksym

Department of Physics and Astronomy, University of Leicester, University Road, Leicester, LE1 7RH, UK.

E-mail: `mr6@le.ac.uk`

Abstract. The 3 and 4-electron states of a gated semiconducting carbon nanotube quantum dot are calculated by exact diagonalisation of a modified effective mass Hamiltonian. A typical nanotube quantum dot is examined and the few-electron states are Wigner molecule-like. The exact diagonalisation method and the rate of convergence of the calculation are discussed.

1. Introduction

In a gated semiconductor nanotube (SNT) quantum dot, electrons or holes are confined electrostatically. The occupancy of this type of dot can be precisely controlled and, importantly, a semiconducting nanotube dot can be completely emptied of electrons. This means that SNT quantum dots are excellent systems in which to study the physics of a few interacting particles.

Like traditional semiconductor quantum dots, SNT dots behave like artificial atoms but there are some differences that make semiconducting carbon nanotube dots unique. First, the SNT bandstructure is unusual, primarily because nanotubes are rolled graphene sheets and the nanotube bandstructure is derived from that of graphene. Second, the geometry of SNT dots is also unusual: the nanotube length is very much greater than its diameter and so SNT dots are quasi-1 dimensional objects. In Ref. [1] we demonstrated that the electronic correlation was large in a large proportion of all physically accessible SNT quantum dots. Taken together with the tube geometry, this means that SNT dots are ideal systems in which to observe quasi-1D Wigner (or all-electron) molecules.

In our approach, we describe the few particle states in a SNT quantum dot with a 1 dimensional, 2 band, effective mass Hamiltonian [1]. However, because the electron correlation is strong and the few-particle states are Wigner molecule-like, care must be taken in the calculation of the electron states: exact diagonalisation, or configuration interaction (CI), techniques must be used to include the correlation effects correctly. In this paper we give an overview of our effective mass description of SNT dots and then discuss the exact diagonalisation scheme in detail. We examine the convergence of the few-electron states and discuss a way of truncating the full CI basis which improves the convergence of the few-electron energy.

© 2010 IOP Publishing Ltd

2. Effective mass description of a nanotube dot

Our dot model is based on the type of semiconducting nanotube quantum dot that is typically studied experimentally [2]. In this type of device, a long ($\sim \mu$m) nanotube is contacted between source and drain electrodes and separated from a back-gate by a thin insulating layer. The quantum dot potential is generated electrostatically by applying a voltage to the back-gate and, in our model, this confinement potential varies parabolically along the length of the tube.

The confined states of a typical SNT dot are broad and may extend over a few hundred nanometres (of the order of a few hundred thousand atoms). However, although the states are broad, their extent is much less than the typical source-drain separation and so we assume our dot is completely isolated. The energy scale of a SNT dot is typically of the order of a few to ten meV, much less than the typical SNT band gap ($\sim$ 1 eV). So, given that the dot length scale is also much larger than the SNT unit cell length, the few-particle states in the dot can be described with an effective mass theory. However, there are new features which must be added to the theory to take account of the unusual nanotube bandstructure.

Nanotubes are rolled graphene sheets, and in each SNT there are two equivalent conduction bands with minima equally spaced either side of $k = 0$. The two conduction bands arise from the equivalence of the K and K' points in the graphene Brillouin zone and there is an additional complication because the dispersion near to each of these points in the graphene zone is linear. In Ref. [1] we demonstrated how to deal with both these effects by expanding the SNT dot states in terms of the two-component Bloch wavefunctions of graphene. The net result is a 1D, 2-band effective mass Hamiltonian with a modified Coulomb interaction, U, that contains all of the effects of the nanotube bandstructure and the details of the nanotube geometry,

$$H = \sum_{i=1}^{N} \left[\frac{-\hbar^2}{2m^*} \frac{d^2}{dz_i^2} + \frac{m^* \omega^2}{2} z_i^2 \right] + \frac{\rho/R}{2\epsilon_r} \sum_{i \neq j} U(z_i - z_j, R). \tag{1}$$

Here N is the number of electrons, $\hbar\omega$ is the parabolic confinement energy, R is the SNT radius, ϵ_r is the dielectric constant, m^* is the effective mass, $\rho = e^2/(4\pi\epsilon_o)$ and

$$U(z, R) = \frac{1}{4}(2u(z, R) + u(z - 3R, R) + u(z + 3R, R)). \tag{2}$$

$u(z, R)$ is derived from the standard form of the interaction between charged particles on the surface of a cylinder,

$$u(z, R) = \frac{2R}{\pi\sqrt{z^2 + 4R^2}} \mathbf{K} \left(\frac{2R}{\sqrt{z^2 + 4R^2}} \right), \tag{3}$$

and $\mathbf{K}$ is an elliptic integral of the first kind which diverges like $\log(z/R)$ as $z \to 0$.

Eq. (2) gives the *intraband* effective interaction. We ignore the *interband* interaction because its effect is small [3, 4]. In this case the band index acts like a pseudospin and our Hamiltonian is block diagonal in both the total electron spin and the pseudospin.

3. Exact diagonalisation

To calculate the N particle energies and wavefunctions we must solve the Schrödinger equation, $H\Psi(z_1 \ldots z_N) = E\Psi(z_1 \ldots z_N)$. We expand the N particle states, $\Psi(z_1 \ldots z_N)$, as a sum of Slater determinants, $\Phi_i(z_1 \ldots z_N)$, constructed from the single particle states of the system,

$$\Psi(z_1 \ldots z_N) = \sum_i a_i \Phi_i(z_1 \ldots z_N), \tag{4}$$

where each $\Phi_i(z_1 \ldots z_n)$ is a normalised eigenfunction of the total spin, S and its z component S_z. In principle, this expansion is exact. However, in reality, the basis set can never be truly complete and must be truncated. In our calculation we use a slight variation of the full configuration interaction scheme in which *all* the possible determinants that can be generated from a given finite single particle basis set are included in the expansion. The size of our finite basis is then determined by a single parameter, a maximum quantum number n, and we can systematically converge our calculation of the few particle states simply by increasing n.

To solve the Schrödinger equation numerically we first transform it into a standard matrix eigenvalue equation by left multiplying with one of the determinants from the expansion in Eq. (4) and integrating over all the $z_1 \dots z_N$ spatial coordinates. We have

$$\sum_i \left(\langle \Phi_j | H | \Phi_i \rangle - E\delta_{ij} \right) a_i = 0,$$

or, in matrix notation,

$$(\mathsf{H} - E\mathsf{I})\, \mathbf{a} = 0,$$

where I is the unit matrix. To find the N particle energies and the expansion coefficients, a_i, we diagonalise this Hamiltonian matrix numerically. The calculation time scales as the size of the matrix cubed and, to reduce this time, we block the Hamiltonian matrix according to the total pseudospin and the symmetry of each N particle state along the nanotube.

To reduce the size of the calculation we also slightly modify the full CI scheme. Our basis functions span a multi-dimensional Hilbert space, but we find that states from the 'corners' of this space do not contribute significantly to the few-electron wavefunctions. So, if

$$\Phi_i(z_1 \dots z_N) = \frac{1}{\sqrt{N!}} \begin{vmatrix} \phi_{m_{i1}}(z_1) & \phi_{m_{i2}}(z_1) & \cdots & \phi_{m_{iN}}(z_1) \\ \phi_{m_{i1}}(z_2) & \phi_{m_{i2}}(z_2) & \cdots & \phi_{m_{iN}}(z_2) \\ \vdots & \vdots & & \vdots \\ \phi_{m_{i1}}(z_N) & \phi_{m_{i2}}(z_N) & \cdots & \phi_{m_{iN}}(z_N) \end{vmatrix}, \tag{5}$$

we find a_i to be negligible if all the m_{ij} quantum numbers are large. We therefore neglect determinants from these 'corners' and use a 'spherical cut off' to truncate the basis when $\left(\sum_{j=1}^{N} m_{ij}^2\right)\right)^{1/2} > n$, as opposed to the usual 'cuboid cut-off' from the full CI scheme in which the basis is truncated when $m_{ij} > n$. The total number of determinants in our expansion at a given maximum quantum number, $M(n)$, depends on the truncation scheme, and we compare the convergence of the ground state energy in each of these two schemes in section 4.

4. Results and discussion

In Ref. [1] we found that the correlation was large in the majority of NT dots, and that all 2-electron states were Wigner molecule-like when $\frac{m^*}{\hbar^2}\left(\frac{(\rho/\epsilon_r)^2}{\hbar\omega}\right) \geq 2.8$. In this work, we therefore focus on a typical strongly correlated SNT dot ($\hbar\omega = 5$ meV, $\epsilon_r = 2.0$) and we examine the 3 and 4 electron states. We choose a 35,0 nanotube with $R = 1.37$ nm and $m^* = 0.034$ as this is the type of tube studied experimentally in Ref. [2]. However, within the effective mass approximation, the details of the chirality are unimportant and our results will be similar for any SNT with similar R and m^*. In the 35,0 NT dot the N-electron ground state energies, E_N, (converged to within 0.1 meV) are $E_3 = 76.4$ meV and $E_4 = 145.1$ meV. Fig. 1 shows the 3 and 4 electron density as a function of z. The states are clearly Wigner molecule-like with a strong peak in the density corresponding to each electron: each electron is almost localised on its own lattice site and the probability of tunnelling between lattice sites is small.

To investigate the convergence of the 3 and 4-electron ground state energy and to compare the 'spherical' and 'cuboid' schemes for truncating the basis set we calculate E_N as a function of n, the maximum single particle quantum number that we include in our basis set. We assume

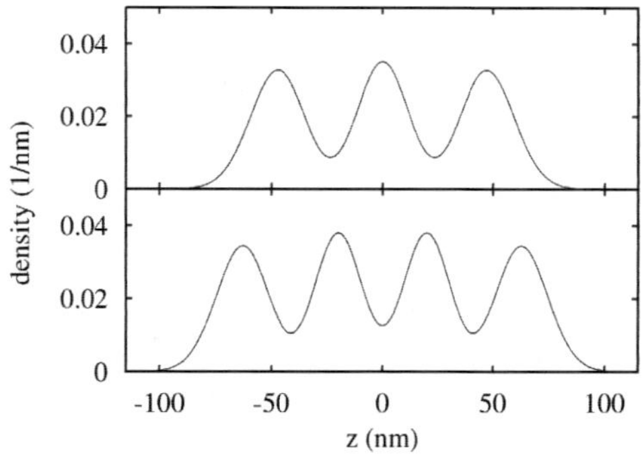

Figure 1. Ground state few-electron densities in a (35,0) SNT quantum dot ($R = 1.37$ nm, $m^* = 0.034$, $\hbar\omega = 5.0$ meV, $\epsilon_r = 2.0$). Top: 3 electron number density. Bottom: 4 electron number density.

that the lowest energy we obtain in our calculations (using the largest number of basis states, M_{max}) is converged. Then $\Delta E_N = E_N(M(n)) - E_N(M_{max})$ gives a measure of the convergence of the ground state energy for each n.

In Fig. 2 we plot ΔE_N as a function of $M(n)$. It is immediately clear that a large number of determinants are needed to give an accurate value for E_N and this is another indication that the electron correlation in SNT dots is important. For example, to converge the ground state energies to within 0.1 meV we need roughly $M = 150$ Slater determinants for 3 electrons and about $M = 3000$ for 4 electrons. It is also apparent that truncating the CI basis with the

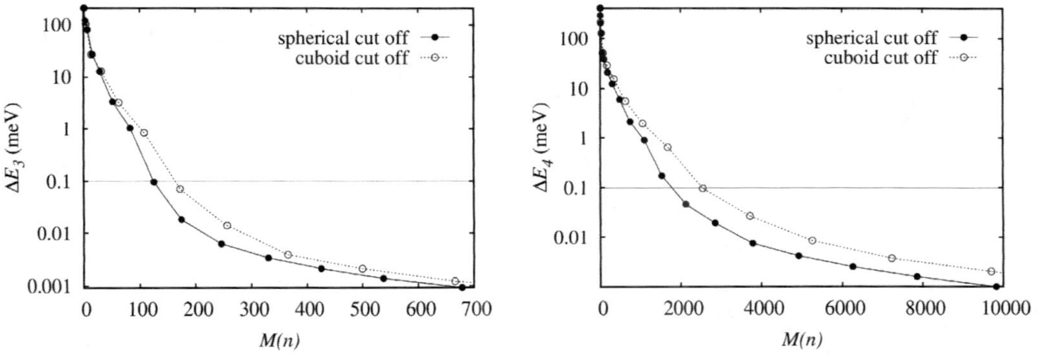

Figure 2. Convergence of the N-electron ground state energy as a function of the number of Slater determinants in the expansion. Left: 3 electrons. Right: 4 electrons. Each point indicates a different value of n in the calculation.

spherical cut off gives an advantage over the full CI method, and this advantage increases with increasing N. For example, to obtain $\Delta E_N = 0.1$ meV with 3 electrons we need $M = 124$ Slater determinants with the spherical cut-off and $M = 164$ with the cuboid cut off. So, using the spherical cut off reduces the computer time by a factor of roughly $(124/164)^3 = 0.57$. Similarly, for 4-electrons the calculation time is reduced by a factor $(1815/2568)^3 = 0.35$.

In conclusion, we have outlined our effective mass description of the N-electron states in SNT quantum dots and demonstrated that many Slater determinants are needed in an expansion of the states to get accurate ground state energies. Also, significant advantages can be obtained using a 'spherical' quantum number cut off to truncate the basis set.

References

[1] Roy M and Maksym P A 2009 *EPL* **86** 37001
[2] Jarillo-Herrero P, Sapmaz S, Dekker C, Kouwenhoven L P and van der Zant H S J 2004 *Nature* **429** 389
[3] Roy M and Maksym P A 2008 *Physica E* **40**, 1616
[4] Secchi A and Rontani M 2009 *Phys. Rev. B* **80** 041404(R)

Journal of Physics: Conference Series **242** (2010) 012008 doi:10.1088/1742-6596/242/1/012008

A quantum transport model for the double-barrier nonmagnetic spin filter

G Isić[1,2], D Indjin[1], V Milanović[3], J Radovanović[3], Z Ikonić[1] and P Harrison[1]

[1] School of Electronic and Electrical Engineering, University of Leeds, LS2 9JT, UK
[2] Institute of Physics, Pregrevica 118, 11080 Belgrade, Serbia
[3] School of Electrical Engineering, University of Belgrade, Bulevar kralja Aleksandra 73, 11120 Belgrade, Serbia

E-mail: elgi@leeds.ac.uk

Abstract. A model for calculating the current and spin polarization in a double-barrier InGaAs resonant tunnelling structure is described with the aim to account for phase-breaking scattering. It is based on the nonequilibrium Green's function method with both elastic and inelastic (LO-phonon) scattering described within the self-consistent first Born approximation. It has been found that the maximum current spin polarization of around 0.4 in the ballistic limit decreases to around 0.1 for scattering transport with scattering-induced broadening of quasi-bound states of around 4meV.

1. Introduction

Recently, it has been suggested that resonant tunnelling structures (RTS) [1] based on III-V semiconductor heterostructures are a promising candidate for a nonmagnetic spin filter, [2, 3, 4, 5]. The mechanism is based on electron tunnelling through quasibound states which are spin-split due to the spin-orbit interaction, manifested as Rashba [6] and Dresselhaus [7] effects.

Numerical estimates of the attainable spin polarization reported so far ignored the detrimental effects of carrier scattering. Here we use the method of nonequilibrium Green's functions [8] adapted for modeling quantum transport in nanodevices, [9, 10, 11].

2. The model

The spin-dependent transport through the InGaAs double-barrier resonant tunneling structure (RTS) shown in Figure 1 is a consequence of the Rashba spin-orbit interaction (RSOI)

$$H_R = \alpha(k_x\sigma_x - k_y\sigma_y), \tag{1}$$

where k_x and k_y are the components of the lateral (perpendicular to the growth direction, denoted as the z axis) wave vector $\mathbf{k}$. σ_x, σ_y are the Pauli matrices and α is the Rashba parameter [12]

$$\alpha = eE_{\text{ext}}\frac{\hbar^2}{2m}\frac{(2E_g+\Delta_{SO})\Delta_{SO}}{E_g(E_g+\Delta_{SO})(3E_g+2\Delta_{SO})}, \tag{2}$$

E_{ext} being the externally applied electric field while $m = 0.023m_0$, $E_g = 356$meV and $\Delta_{SO} = 410$meV are the InAs band parameters. RSOI creates the energy difference $\Delta E(|\mathbf{k}|) = 2\alpha|\mathbf{k}|$

© 2010 IOP Publishing Ltd

between the two resonant energy levels denoted by E_R^+ and E_R^- in Figure 1. The value of α

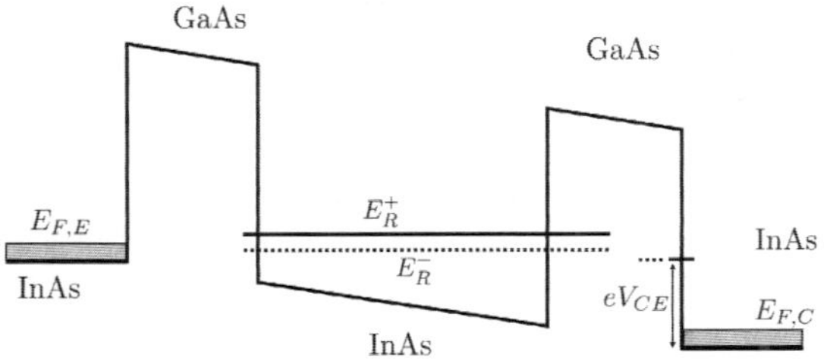

Figure 1. The effective electron potential profile in a biased InGaAs RTS. The conduction band offset is taken to be $\Delta E_c =$ 792meV, $m_{\mathrm{GaAs}} = 0.067m_0$, the barrier and well width $L_{\mathrm{b}} =$ 3nm and $L_{\mathrm{w}} =$ 6nm, respectively, as in [2] while the Fermi energies in the contacts are $E_{F,E} = E_{F,C} =$ 20meV.

obtained from (2) for $E_{ext} = 20\mathrm{mV/nm}$ (corresponding to the current peak in the $J(V_{CE})$ curves shown in Figure 2) has an order-of-magnitude agreement with values found experimentally in [13]. For $k = k_F$ ($E_F =$ 20meV) it yields $\Delta E(k_F) =$ 2meV. Since the spin splitting has the same order of magnitude as the scattering-induced broadening of the resonant energy levels [1], we conclude that to model the spin polarization in a RTD, one has to go beyond the ballistic transport picture.

The quantities describing the electronic system are the greater ($G^>(E)$) and lesser ($G^<(E)$), Green's functions. Once $G^>(E)$ and $G^<(E)$ are found, it is straightforward to obtain the electron concentration, density of states and current density, [11]. The interaction with InAs optical phonons ($\omega_{LO} =$ 29meV) and elastic phase breaking are calculated within the self-consistent first Born approximation, the procedure for which we briefly outline here.

We start by finding the retarded ($\Sigma^R(E)$), lesser ($\Sigma^<(E)$) and greater ($\Sigma^>(E)$) self-energies. They are obtained by adding up the contributions of various interactions. In particular, we have

$$\Sigma^\gamma(E) = \Sigma_E^\gamma(E) + \Sigma_C^\gamma(E) + \Sigma_{\mathrm{ph}}^\gamma(E) + \Sigma_{\mathrm{el}}^\gamma(E), \quad \gamma = R, <, > . \tag{3}$$

The indices E and C refer to terms due to interaction with emitter and collector reservoirs, respectively, [11]. $\Sigma_{\mathrm{ph}}^\gamma(E)$ and $\Sigma_{\mathrm{el}}^\gamma(E)$ describe the LO phonon and elastic phase breaking. In the first iteration, these two are assumed zero. At the end of each iteration, new values for the scattering self-energies are obtained and compared with previous ones. Typically, 10 to 30 iterations were required to obtain a relative error below 10^{-5}.

Then, we find $G^R(E)$ according to

$$G^R(\sigma, |\mathbf{k}|, z, z'; E) = \Big(E - H(\sigma, |\mathbf{k}|) - \Sigma^R(z, z'; E)\Big)^{-1}, \tag{4}$$

$$H(\sigma, |\mathbf{k}|) = -\frac{\hbar^2}{2m}\frac{d^2}{dz^2} + \frac{\hbar^2 k^2}{2m} + \sigma\alpha|\mathbf{k}| + V(z). \tag{5}$$

$V(z)$ comprises the band-edge potential and the linear term due to the external field, see Figure 1. $\sigma = \pm$ denotes the spin subband. For calculations we use the $|\mathbf{k}, z\rangle$ basis with N grid points for the z axis and the finite difference approximation. Thus, for fixed values of σ, $|\mathbf{k}|$ and E, $G^R(\sigma, |\mathbf{k}|, z, z'; E)$ becomes a $N \times N$ matrix. The lesser and greater Green's functions are found from the kinetic equation

$$G^\lambda(\sigma, |\mathbf{k}|, z, z'; E) = \int dz_1 \int dz_2 G^R(\sigma, |\mathbf{k}|, z, z_1; E)\Sigma^\lambda(z_1, z_2; E) G^A(\sigma, |\mathbf{k}|, z_2, z'; E), \tag{6}$$

where $G^A(E)$ is the Hermitian conjugate of $G^R(E)$ and the z-integration goes from the emitter to the collector contact (left and right end of the structure shown in Figure 1).

Journal of Physics: Conference Series **242** (2010) 012008 doi:10.1088/1742-6596/242/1/012008

New values of the lesser and greater scattering self-energies are given by the first Born approximation ($\varphi =$'ph','el')

$$\Sigma_{\varphi}^{\lambda}(z,z';E) = \frac{U_{\varphi}^2}{8\pi^2}\delta(z-z') \int d\mathbf{k} \left(G^{\lambda}(+,|\mathbf{k}|,z,z;E \pm \hbar\omega_{\varphi}) + G^{\lambda}(-,|\mathbf{k}|,z,z;E \pm \hbar\omega_{\varphi}) \right) \quad (7)$$

$\pm$ in front of the phonon energy $\hbar\omega_{\varphi}$ means + for $\lambda =<$ (electron relaxation from $E+\hbar\omega_{\varphi}$ to E) and $-$ for $\lambda =>$ (hole relaxation from $E-\hbar\omega_{\varphi}$ to E). The above expression accounts only for spontaneous phonon emission, which is appropriate for low temperatures such that the average number of phonons $n_B(\hbar\omega_{LO}) = (\exp(\hbar\omega_{LO}/kT)-1)^{-1}$ is much smaller than 1. For elastic scattering, we put $\omega_{\rm el} = 0$.

Finally, the imaginary part of the retarded scattering self-energy is given by.

$$\Sigma_{\varphi}^{R}(z,z';E) = \frac{1}{2}\left(\Sigma_{\varphi}^{>}(z,z';E) - \Sigma_{\varphi}^{<}(z,z';E)\right), \quad (8)$$

while the real part of $\Sigma^R(E)$ is neglected since it merely shifts the energy levels in the system. The self-energies found from (7) and (8) are local ($\sim \delta(z-z')$) due to the assumption that the coupling to scatterers is independent on $\mathbf{k}$. It is exact for the deformation potential coupling to optical phonons, but not for polar coupling in which case the effective local scattering strength U^2 can be found by averaging, see Appendix C of [14]. In this work $U_{\rm ph}^2 = 2500{\rm meV\AA}/D_{\rm InAs}$ and $U_{\rm el}^2 = 250{\rm meV\AA}/D_{\rm InAs}$ with $D_{\rm InAs} = m/2\pi\hbar^2$ (local density of states of a spin subband in a InAs two-dimensional electron gas) have been used. These gave average elastic and phonon scattering rates of around $10^{12}s^{-1}$ at the current main peak shown in Figure 2 inset ($V_{CE} \approx 270$mV) and the enhanced value for phonon scattering of around $10^{13}s^{-1}$ at the phonon peak ($V_{CE} \approx 320$mV), which is comparable to values reported in Ref. [10]. For $V_{CE} \approx 270$mV it has been found that the quasi-bound state broadening is 0.8meV in the ballistic limit, 4.5meV with only elastic scattering and 4.9meV when both elastic and inelastic scattering are taken into account. Thus the scattering-induced broadening of quasi-bound states found using the above scattering strengths is consistent with typical values found in the literature [1].

The energy resolved spin-subband current density entering the system through contact $\gamma = E, C$ (emmiter and collector) is given by [11]

$$i_{\gamma}(\sigma,E) = \frac{e}{4\pi^2 h}\int dz \int dz' \int d\mathbf{k} \left(\Sigma_{\gamma}^{<}(z,z';E)G^{>}(\sigma,|\mathbf{k}|,z',z;E)\right.$$
$$\left. - \Sigma_{\gamma}^{>}(z,z';E)G^{<}(\sigma,|\mathbf{k}|,z',z;E)\right). \quad (9)$$

The total spin-subband current densities flowing through the device are obtained by integrating $i_{\gamma}(\sigma,E)$ over energy

$$J(\sigma) = \int dE i_E(\sigma,E) = -\int dE i_C(\sigma,E), \quad (10)$$

while the total current density is $J = J(+) + J(-)$. In absence of a lateral electric field, the Kramers degeneracy implies a zero spin polarization of the total current. Thus, the one-sided collector geometry is assumed, [4], in which case the spin polarization is given by

$$P = \frac{2}{\pi}\frac{J(+)-J(-)}{J(+)+J(-)}. \quad (11)$$

3. Results

Figure 2 shows the dependence of P and current density J (inset) on the external bias. For reasons of compactness, curves for $J(+)$ and $J(-)$ separately are not shown. They have a shape similar to J (multiplied by a factor of 0.5) and are mutually displaced in voltage by approximately $\Delta V_{CE} = 2\text{mV} \sim \Delta E(k_F)/e$. Maxima in P occur at places where there is a sudden change of the magnitude of either $J(+)$ or $J(-)$: around the current turn-on ($V_{CE}^{\text{on}} \approx 245\text{mV}$) the $J(-)$ component suddenly rises leading to a negative peak in P; the positive peak in P occurs at the current turn-off voltage ($V_{CE}^{\text{off}} \approx 285\text{mV}$) where $J(-)$ goes to zero before $J(+)$. The polarization is reduced by the broadening of features in $J(V_{CE})$, which is the main effect of carrier scattering. The maximal value of P for the fully ballistic transport ($U_{\text{ph}} = 0$, $U_{\text{el}} = 0$) of around 0.4 is reduced to around 0.1 when elastic scattering is present. LO-phonon scattering causes the phonon-peak at $V_{CE} \approx 320\text{mV}$ in both subband currents, but it is so wide that no feature in $P(V_{CE})$ is observed. Compared to the case when only the elastic scattering is taken into account ($U_{\text{ph}} = 0$, $U_{\text{el}} \neq 0$), the case with LO-phonon scattering does not create a significant difference in $P(V_{CE})$ because when V_{CE} is large enough to make the resonant transmission with phonon emission possible, there are no carriers that can be resonantly transmitted (without scattering) hence no sharp features in $J(V_{CE})$ exist.

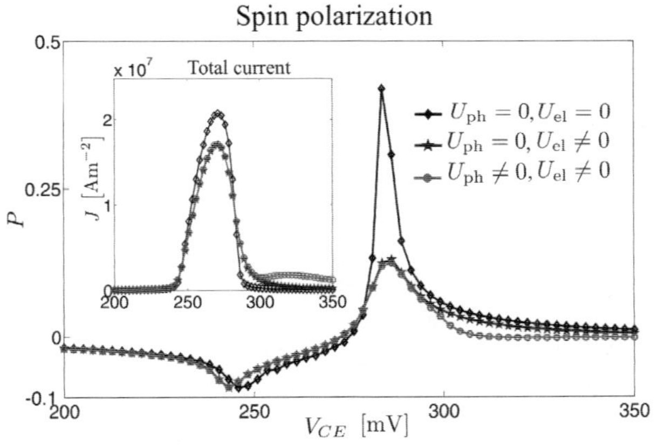

Figure 2. Calculated values of the spin polarization and current density (inset) for the InGaAs RTS described in Figure 1 at $T = 4.2K$. Three cases have been considered: ballistic transport (diamonds), elastic scattering only (stars) and both elastic and LO-phonon scattering (circles).

In summary, we have described a model to account for effects of carrier scattering on spin polarization of the current. By choosing numerical values of scattering strengths that yield average relaxation times of around 1ps (and around 0.1ps for resonant LO-phonon emission), the peak in spin polarization is found to decrease from 0.4 in the ballistic case to around 0.1 when carrier scattering is present.

Acknowledgments

Authors acknowledge the support of NATO Collaborative Linkage Grant (reference CBP.EAP.CLG 983316). G. I. is grateful for the support from ORSAS (UK), University of Leeds, the School of Electronic and Electrical Engineering and the Serbian Ministry of Science project No. 141047. V. M. and J. R. acknowledge the support from the Serbian Ministry of Science project No. 141006.

References

[1] Mizuta H and Tanoue T 1995 *The Physics and Applications of Resonant Tunnelling Diodes* (Cambridge: Cambridge University Press)
[2] Voskoboynikov A, Lin S S, Lee C P and Tretyak O 2000 *J. Appl. Phys.* **87** 387
[3] Glazov M M, Alekseev P S, Odnoblyudov M A, Chistyakov V M, Tarasenko S A and Yassievich I N 2005 *Phys. Rev.* B **71** 155313

[4] Koga T, Nitta J, Takayanagi H and Datta S 2002 *Phys. Rev. Lett.* **88** 126601
[5] Isic G, Radovanovic J and Milanovic V 2007 *J. Appl. Phys.* **102** 123704
[6] Bychkov Y A and Rashba E I 1984 *J. Phys.* C **17** 6039
[7] Dresselhaus G 1955 *Phys. Rev.* **100** 580
[8] Keldysh L V 1965 *Sov. Phys.* JETP **20** 1018
[9] Datta S 1990 *J. Phys.: Condens. Matter* **2** 8023
[10] Lake R and Datta S 1992 *Phys. Rev.* B **45** 6670
[11] Datta S 1999 *Electronic Transport in Mesoscopic Systems* (Cambridge: Cambridge University Press)
[12] de Andrada e Silva E A, La Rocca G C and Bassani F 1997 *Phys. Rev.* B **55** 16293
[13] Grundler D 2000 *Phys. Rev. Lett.* **84** 6074
[14] Lake R, Klimeck G and Datta S 1993 *Phys. Rev.* B **47** 6427

Journal of Physics: Conference Series **242** (2010) 012009 doi:10.1088/1742-6596/242/1/012009

Predictive Microscopic Approach to Transport in THz Quantum Cascade Lasers

T. Schmielau and M.F. Pereira

Materials and Engineering Research Institute, Sheffield Hallam University

S1 1WB, Sheffield, United Kingdom

mpereira@shu.ac.uk

Abstract. In this paper we present further details of a predictive Keldysh nonequilibrium many body Green's functions theory for quantum transport including high order electron-electron, electron-phonon, electron-impurity and interface roughness scattering processes. Our approach is fully frequency and momentum dependent. Local current conservation is observed even if only few states are considered. The resulting algorithm has led to good agreement with experimental observations.

1. Introduction

The Quantum Cascade Laser (QCL) is the best candidate for efficient generation of terahertz (THz) radiation. QCL development is one of the current frontiers in semiconductor science from both the fundamental and applications points of view [1]. The systematic exploitation of THz waves has been hindered by the lack of a compact coherent source providing high output power. There is a huge potential for THz technology in a varied list of applications; detecting tumours and skin cancers, pharmaceutical applications, detecting and discriminating explosive threat materials, environmental sensing and gas monitoring, industrial process control, astronomy, semiconductor imaging, security and medical imaging and telecommunications [2]. A number of different materials and systems have been proposed to create compact sources [3] but so far QCLs remain the best candidates for this role.

It cannot be over emphasised what a tour de force a QCL is in both the design and the growth. For example, an average THz QCL contains about 1000 layers and takes about eight hours to grow, as compared to the two hours required for a moderately complex commercial device such as a GaAs high electron mobility transistor. Each design has to be optimised for the wavelength chosen, materials used, and growth technique employed, and trade-offs have to be made for pulsed or continuous use, and operating temperature. In fact, it is questionable how often this ideal is actually reached as design and growth are so difficult. The main difficulty so far is to achieve population inversion and it may well be that room temperature operation will only be reached through innovative concepts, like lasing without inversion [4-6]. A fully predictive and user-friendly QCL simulator is needed by manufacturers and research teams to design and simulate new devices easily and solve the current technical problems faster. In this paper we complement the material published recently [7-9] on our predictive transport simulator by presenting numerical results for the energetically and spatially resolved electronic density and density of states in a THz QCL. We further give more details of the performance of our algorithms.

© 2010 IOP Publishing Ltd

2. Summary of the Main Equations and Numerical Method

Electronic states in the QCL are described by in-plane waves with wavevector **k** = (k_x, ky) and (Wannier or position) states α,β... in growth direction (z) and are denoted, e.g. $\Psi_\alpha(z)$.We define $\alpha = (\mu,m,s)$ and $\beta = (\nu,n,s')$ and where m; n indicate the basic period where the state is located and the subband indices, ν,μ number the states inside that period. The spin indices s,s' = ± 1/2 are dropped from now on as all occupations and interactions are spin independent. As translational invariance is broken in the growth direction, the usual Dyson equation for the retarded G becomes a matrix equation in the states

$$G^{ret}_{\alpha\beta}(k,E) = (G^{ret}_0(k,E) - \Sigma^{\mathrm{ret}}(\mathrm{k,E}))^{-1}_{\alpha\beta}, \tag{1}$$

where the inverse unperturbed retarded Green's function $G^{ret}_{0\alpha\beta}(k,E)^{-1} = E\delta_{\alpha\beta} - H^0_{\alpha\beta} + i\varepsilon$ is determined by the single-particle Hamiltonian $H^0 = H_{kin} + V_{SL} + V_H + V_{field}$ containing kinetic energy, superlattice potential, mean field, and electric potential $V_{field} = e\varepsilon z$ due to the applied electric field ε, respectively. In order to reduce the matrix equation to finite size, the periodicity of the superlattice is exploited (period length d)

$$G_{\mu m+p,\nu n+p}(k,E+e\varepsilon dp)=G_{\mu m,\nu n}(k,E). \tag{2}$$

Furthermore, Greens Functions spanning more than two basic periods are assumed to vanish. The greater (>) and lesser (<) components of the GF, which determine the actual occupation of the electronic states, are then obtained from the Keldysh relation

$$G^<_{\alpha\beta}(k,E) = \sum_{\mu\nu} G^{ret}_{\alpha\mu}(k,E)\cdot\Sigma^<_{\mu\nu}(k,E)\cdot\left(G^{ret}_{\beta\mu}(k,E)\right)^*,$$
(3)

and similarly exchanging (<) by (>). The additional initial condition term has been omitted from Eq. 3 as it vanishes due to dissipation. As a numerical cross-check it has been verified that the Kubo-Martin-Schwinger (KMS) [10] condition between $G^>$ and $G^<$ is fulfilled at zero applied voltage, indicating relaxation to thermal equilibrium. The spatially and energetically resolved electron density reads (see also Ref. [11])

$$n_e(z,E) = \sum_{\vec{k}} n_e(z,K,E) = \frac{1}{i\pi A}\sum_{\alpha\beta\vec{k}} \Psi^*_\alpha(z)\Psi_\beta(z)G^<_{\beta\alpha}(k,E), \tag{4}$$

where A denotes the sample area.

3. Numerical Results

Wavenumber resolved quantities can be defined by leaving out the k summation as shown in Eq. (4).Figures 1 and 2 depict the electronic density resolved both spatially and energetically at k=0 for the QCL of Ref. [12]. Equilibrium conditions are shown in Fig. 1 while in Fig. 2 the system is out of equilibrium and polarized with a voltage drop of 62 mV per period. The development of gain requires a combination of allowed transitions, upper laser states more populated than the lower sates (population inversion) and a relatively small dephasing so that the resulting gain can overcome waveguiding losses. Local gain is achieved when certain regions of k-space (normally for small k) are inverted and compensate the losses in other k-space regions when all contributions are summed up. The active region is typically designed so that electronic overlap between the relevant states lead to large transition dipole moments. Figure 1 shows only lower laser levels occupied at equilibrium while Fig.2 under nonequilibrium conditions clearly shows more occupation in the upper levels in the active region 5.5nm < z < 22.4nm. This indicates local population inversion in k-space that can potentially lead to gain and thus lasing.

Those plots allow further insight into the nonequilibrium charge distribution in the structure and use it to analyse potential design failures and re-design the lasers based on those studies. Similar studies already led to deeper insight into the characteristics of mid infrared and THz QCLs [11, 13]. The self energies included in the calculations include high order electron-electron, electron-phonon, electron-impurity and interface roughness scattering processes. To the best of our knowledge this is the first time that those plots are presented with a consistent inclusion of electron-electron scattering within a dynamic single-plasmon pole approximation. The numerical results presented here were obtained with our optimized simulator, which is fully parallelized and optimized for vector units (SSE2 / AltiVec). The self energy calculation, which is the main numerical task, achieves ~ 60% of the CPU's theoretical peak performance (~ 43 GFLOP/s on a 2.2 GHz quad core CPU).

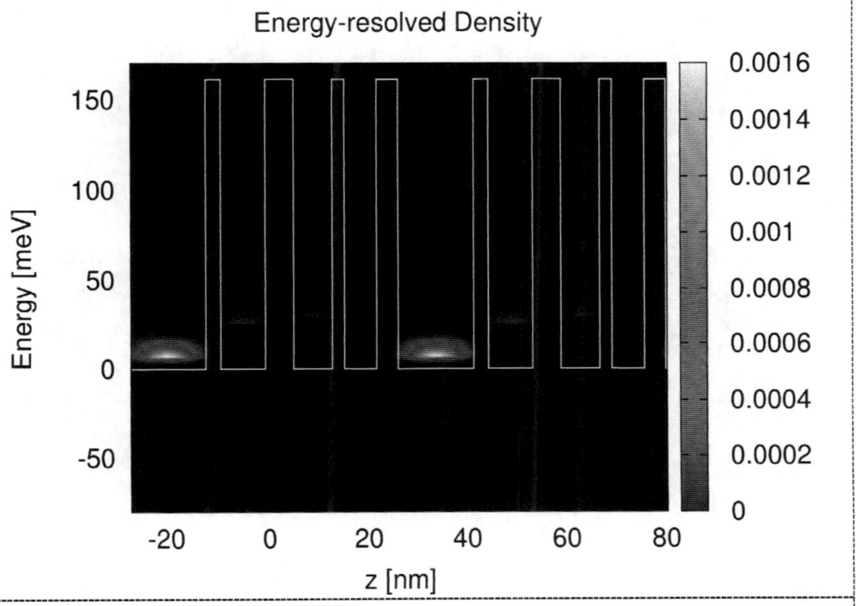

Figure 1. Equilibrium energetically and spatially resolved electronic density at $k=0$, $n_e(z,k=0,E)$ in Eq. 4, for the QCL of Ref. [12].

A compensation between coherent and scattering processes leads to local current conservation even if only a small number of states is considered. Our algorithm is fully frequency and momentum dependent The numerical results presented here assume the plasmons to be in thermal equilibrium with the lattice phonons. This approximation should be more accurate when the plasma frequency characterizing a given subband [14] is close to the optical phonon frequency, thus leading to stronger coupling between those modes [15, 16]. However, the plasmons are still part of the electron gas and in future work we plan to develop a more complete theory that also consider details of the energy transfer from plasmons to phonons, leading to a nonequilibrium plasmon distribution function.

In summary, our efficient numerical scheme can be used to predict transport of electrons as they cascade down a polarized quantum cascade laser and yields microscopic insight that can help understand the current bottlenecks in the development of new THz designs. It can help create structures that will operate at higher temperature more efficiently.

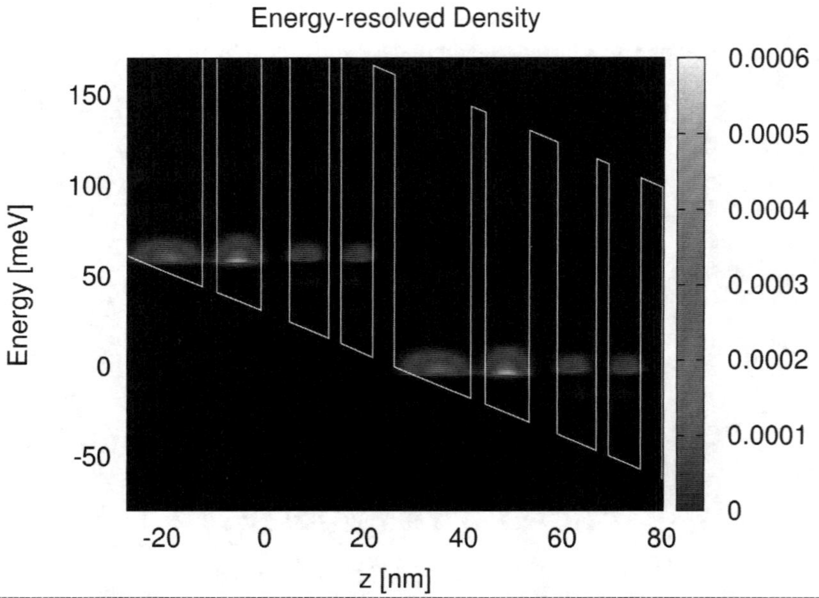

Figure 2. Nonequilibrium energetically and spatially resolved electronic density at *k=0*, $n_e(z,k=0,E)$ in Eq. 4,for the QCL of Ref. [12] polarized with a voltage drop of 62 mV per period. Note the local population inversion in the active region 5.5 nm < z < 22.4 nm

References

[1] R. Köhler, A. Tredicucci, F. Beltram, H. E. Beere, E. H. Linfield, G. A. Davies and D. A. Ritchie, Advances in Solid State Physics 43, 327 (2003).
[2] M.F. Pereira, Journal of Telecommunications and Information Technology 4, 118 (2009).
[3] M.E. Portnoi, O.V. Kibis, M. Rosenau da Costa, Superlattices and Microstructures 43, 399 (2008).
[4] A. Wacker, Nat. Phys. 3, 298 (2007).
[5] R. Terazzi, T. Gresch, M. Giovanni, N. Hoyler, F. Faist, and N. Sekine, Nat. Phys. 3, 329 (2007).
[6] M. F. Pereira Jr., Phys. Rev. B, vol. 78, 245305-1 (2008).
[7] T. Schmielau and M.F. Pereira, Appl. Phys. Lett. 95, 231111 (2009).
[8] T. Schmielau and M.F. Pereira, Microelectronics Journal 40, 869 (2009).
[9] T. Schmielau and M.F. Pereira, physica status solidi b 246, 329 (2009).
[10] R. Zimmermann, Many-Particle Theory of Highly Excited Semiconductors, Teubner, Leipzig (1987).
[11] A. Wacker, phys. stat. sol. (c) 5, No. 1, 215–220 (2008).
[12] S. Kumar, B. S. Williams, S. Kohen, Q. Hu and J. Reno, Appl. Phys. Lett. 84, 2494 (2004).
[13] R. Nelander, A. Wacker, M.F. Pereira Jr., D.G. Revin, M.R.Soulby, L.R. Wilson, J.W. Cockburn, A.B. Krysa, J.S. Roberts, and R.J. Airey, Journal of Applied Physics 102, 113104 (2007).
[14] M. F. Pereira, Jr., W. Chow, and S. W. Koch, J. Opt. Soc. Am. B 10, 765 (1993).
[15] C. Peschke, J. Appl. Phys. 74, 327 (1993).
[16] P. Borowik, J. Appl. Phys. 82, 4350 (1997).

Journal of Physics: Conference Series **242** (2010) 012010 doi:10.1088/1742-6596/242/1/012010

From *ab initio* properties of the Si-SiO_2 interface, to electrical characteristics of metal-oxide-semiconductor devices

S Markov[1,2,4], P Sushko[3], C Fiegna[2], E Sangiorgi[2], A Shluger[3], A Asenov[1]

[1] Dept. of Electronics & Electrical Eng., Rankine Building, Glasgow, G12 8LT, UK

[2] ARCES-IUNET, University of Bologna, via Venezia 52, Cesena 47023, Italy

[3] Dept. of Physics and Astronomy, University College of London, WC1E 6BT, UK

E-mail: s.markov@elec.gla.ac.uk

Abstract. The gradual transition of the band-gap at the Si-SiO2 interface affects quantisation and leakage characteristics of MOS inversion layer. We establish a link between first principles DFT simulations of the interface, and continuum simulations in the effective mass approximation, in order to obtain a realistic description of the band-gap transition for device modelling. The simplistic approach of obtaining real-space-dependent band-gap profile from the *ab initio* calculated electronic structure results in uncertainty of the simulated device characteristics. This uncertainty is small however, when compared to the magnitude of the simulated impact of the transition layer. A linear transition of the band-gap over 6 – 7 Å in the oxide approximates well the effects simulated with the realistic band-gap profile from DFT.

1. Introduction

The change of the atomic structure at the Si-SiO_2 interface leaves additional electronic states, energetically aligned closer to the Si conduction band, in the first 2 – 6 Å of the SiO_2 away from the interface, which implies a non-abrupt transition of the band-gap within the oxide [1]. This transition is of great importance for the characterization of gate oxide (or high permittivity gate stacks) and gate leakage in metal-oxide-semiconductor (MOS) field effect transistors (FETs), since the width of the interface transition (a few Å) is of the order of the oxide thickness itself [2]. Although physical understanding of the involved phenomena could be obtained with the help of *ab initio* modelling methods (e.g. density functional theory), DFT modelling is impractical for device simulations (of model systems of 10^8 atoms, at a finite bias), and even a tight-binding approach presents a limiting computational cost for that purpose [3]. Accordingly, we established a hierarchical simulation approach, allowing DFT-calculated electronic properties of the interface to be incorporated as real-space-dependent macroscopic parameters (e.g. band-gap/offsets) in our device simulations [4]. The translation of *ab initio* electronic structure to macroscopic material parameters is not without ambiguity however, owing to certain freedoms in setting up the DFT simulations (e.g. choice of functional), and the procedure of band-gap determination and scaling. In this paper we explore the extent to which such uncertainties influence the magnitude of the simulated impact on device characteristics, due to the band-gap transition layer.

[4] Corresponding author; SM acknowledges financial support by European Community via "Marie Curie" Fellowship (EDITH: MEST-CT-2004-504195)

© 2010 IOP Publishing Ltd

2. Methodology

We aim to account for certain *microscopic* properties of the Si-SiO_2 interface in *macroscopic*, continuum device simulations. Our hierarchical simulation approach consists of three steps: 1) calculation of the atomic and electronic structure of the oxide interface, based on DFT simulations of a 3D-periodic SiO_2-Si-SiO_2 super-cell, adopting α-quartz model for SiO_2; 2) translation of the electronic structure to continuous, one-dimensional space-dependent conduction and valence band offsets (relative to the conduction band edge of bulk Si); 3) one-dimensional device simulation of a metal/oxide/p-Si(001) structure, using a self-consistent Poisson-Schrödinger (PS) solver. The details of the atomic structure model and optimisation during step 1), and of the PS-solver of step 3), are described elsewhere [4]. Here, we elaborate on step 2).

The starting point for this is the electronic structure calculated for the selected super-cell with CRYSTAL-2003 computer code, using gradient-corrected density functionals and Gaussian-type basis set. The total density of states (DOS) is obtained using the one-electron *E(k)* dispersion, calculated at a pre-defined set of *k*-points in reciprocal space. The corresponding one-electron states are represented as linear combination of atomic orbitals, allowing the projection of the total DOS on such an atomic basis, and therefore, the calculation of the atom-projected DOS (pDOS). Results for pDOS, obtained in this way depend on the chosen density functional and the choice of pre-defined *k*-points. For the super-cell used in our study, we have three sets of pDOS data: SET1 (PBE functional), SET2 (B3LYP functional), SET3 (B3LYP functional, and 2 times denser *k*-grid, compared to other two sets).

Figure 1 shows the conduction and valence band-edge (CBE and VBE) profiles extracted from the pDOS of SET1. The band-edges are determined from the highest occupied and lowest unoccupied states associated with each atom. Different criteria of the minimum pDOS value result in slightly different band-edge profile. For $0.05 < min(pDOS) < 0.1$ (values in a.u.), the extracted profiles nearly overlap. We find the same qualitative picture for the other two sets, and similar profiles presented in other numerical and experimental studies [5]. Note that oxygen atoms within the sub-stoichiometric oxide (delimited by dotted, vertical line in figure 1) contribute states that are energetically in the same range as the ones contributed from Si-atoms. Analogous effect is observed for the valence-band edge.

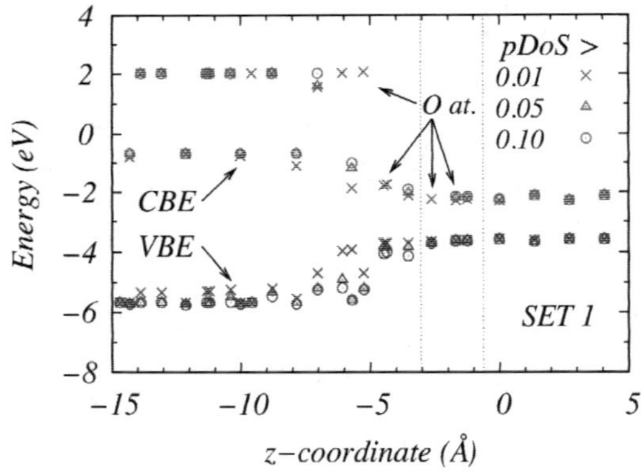

Figure 1. CBE and VBE profiles for SET1, for three band-edge determination criteria, as shown. Results for 0.05<min(pDOS)<0.1 nearly overlap. Si(SiO_2) is along the positive(negative) direction.

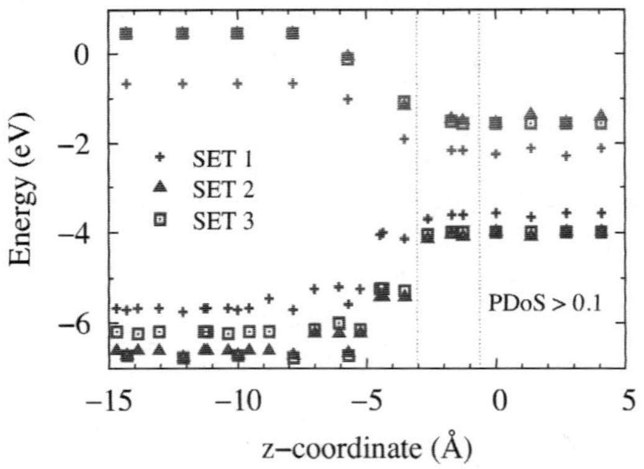

Figure 2. CBE and VBE profiles for the three different sets. The band gap strongly depends on the chosen functional. The number of *k*-points affects mostly SiO_2 VBE.

Figure 2 compares the band-edge profiles extracted for each of the three sets, at the same band-edge determination criterion of (pDOS)>0.1. It highlights a quantitative discrepancy, arising from the use of different functionals. Both PBE and B3LYP underestimate the experimental (bulk) SiO_2 band-gap of approximately 8.9 eV, and overestimate (significantly for the B3LYP sets) the bulk Si band-gap (1.12 eV). The latter effect is partly due to the small thickness of ordered Si (5 Å simulated). The former is attributed to the DFT approach, as applied here, and requires scaling of the band gaps prior to incorporation of band-edge profiles in the device simulator. Figure 2 also shows the effect of a denser *k*-point grid, to flatten the band-edge in the "bulk"-like parts, most notably for the SiO_2 VBE.

For device simulation, not only the band-gap E_G, but also the conduction and valence band offsets ΔE_C, and ΔE_V, are important. As we are interested in the inversion and gate leakage characteristics of a metal/SiO_2/p-Si(100) structure, the penetration of electrons in the oxide becomes the main impact factor, and it depends on the potential barrier for electrons at the interface. Therefore, we choose ΔE_C to be the leading parameter for scaling of the oxide region. The accuracy of the oxide band-gap is relaxed in this case, leading to an overestimation of the valence band offset. This has no effect in our study, since hole density only in the bulk-Si is relevant. For the region of the Si substrate E_G is the leading scaling parameter.

The scaled conduction and valence band-edge profiles, E_C and E_V respectively, are obtained from: $E_C(z) = E_0 + \alpha(z)\Delta E_C^{DFT}(z)$, and $E_V(z) = E_C(z) - \alpha(z)\, E_G^{DFT}(z)$, where E_0 is our energy reference, taken as the CBE of bulk Si. The scaling coefficient $\alpha(z)$ has a different value in SiO_2, and in Si, as discussed above: $\alpha(SiO_2) = \Delta E_C^{NOM}/\Delta E_C^{DFT}$, and $\alpha(Si) = E_G^{NOM}/E_G^{DFT}$, where ΔE_C^{NOM}=3.15 eV, and E_G^{NOM}=1.12 eV. In the sub-stoichiometric oxide, α is assumed to change linearly from the value in Si, to that in SiO_2: $\alpha(SiO_X) = ((\alpha(SiO_2) - \alpha(Si))/t_{SiOx})\, z+\alpha(Si)$, where z is the distance from the interface, and t_{SiOx} is the thickness of the sub-stoichiometric oxide (~2 Å for the α-quartz SiO_2).

Figure 3 shows the CBE and VBE profiles around the oxide interface, for the three data sets, after scaling according to the above equations. Although the three profiles look quite similar, the relative difference between them exceeds twice the thermal potential (~26 meV), in the transition region. In the next section, we show the implication of this discrepancy, on the simulated device characteristics.

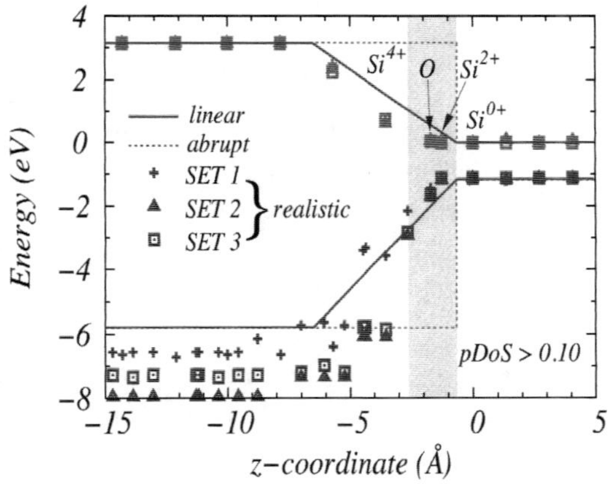

Figure 3. DFT-obtained CBE and VBE profiles, after scaling (symbol), compared with the *linear* approximation and the conventional *abrupt* interface band-gap transition. O-coordination of Si is indicated within and around the suboxide.

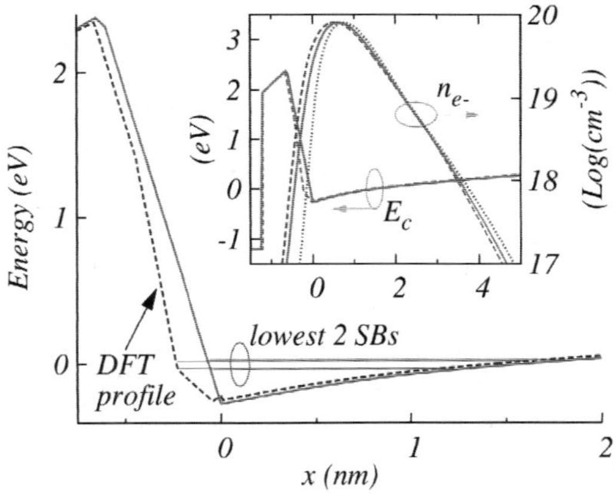

Figure 4. Conduction band edge, and lowest two subbands (SBs) for the *linear* (line) and *realistic* (dash, SET1) interface band-gap profiles. Inset shows the corresponding electron density for *abrupt* (dots), *linear*, and *realistic* case.

3. Device simulation results

We refer to the CBE and VBE profiles obtained as discussed above as *realistic*, despite their quantitative discrepancy between each other. Comparison of these profiles with a linearly approximated band-gap transition is shown in figure 3. The main feature of the *realistic* profiles is the rather slow initial development of the band-gap, followed by a steeper rise. This feature is reported in other DFT and experimental studies of the Si-SiO_2 interface [5], and is in contrast to the linear profile of a similar total transition width (~6 Å).

Figure 4 shows conduction band and electron density (inset) in a MOS structure biased in strong inversion. The different curves reflect the different barrier models from figure 3 (*realistic*, DFT profile corresponds to SET1). The lowest two subbands (hosting almost the entire electron population in the substrate) are energetically close to the bottom of the quantum well. For the *realistic* profile, the well is wider, and the potential barrier is thinner. This induces more inversion charge in proximity to the oxide relative to the simulated *linear* and *abrupt* cases. Therefore, the impact of the non-abrupt band-gap transition is enhanced, due to the non-linear band-edge profile, and the *exact* profile of the band evolution, rather than the total transition width, determines the magnitude of the observed effect.

Figures 5 shows the energy level and occupancy (percentage of the total electron density) of the lowest two subbands, as a function of gate bias, for all simulated interface barrier profiles. The most important observation here is that the uncertainty in the definition of the *realistic* profiles (resulting from SET1 to SET3) is reflected in a spread of the device characteristics (note the group of dashed curves). Quantitatively however, this spread is small compared to the magnitude of the simulated impact, due to band-gap transition. The same conclusion is drawn from figure 6, showing the direct tunnelling gate current simulated for all simulated interface barrier profiles.

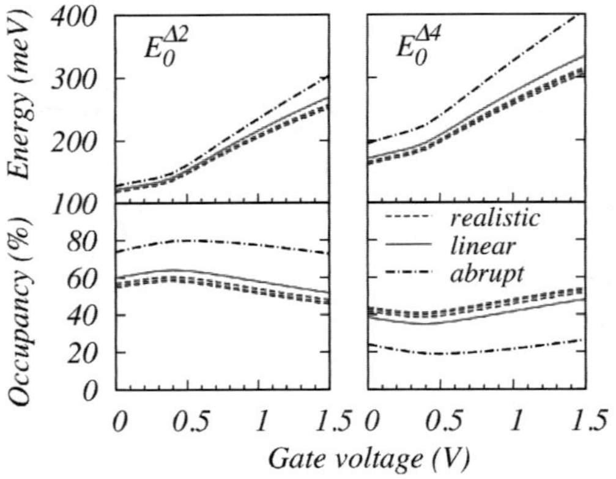

Figure 5. Energy (top) and occupancy (bottom) of the lowest two subbands. The spread of *realistic* curves is small compared to their shift from *abrupt* line.

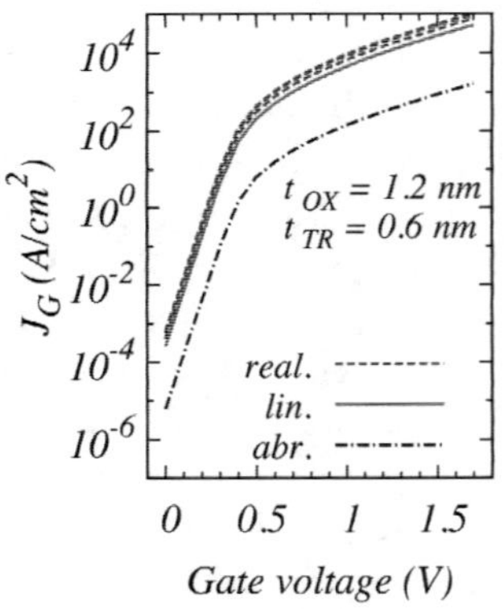

Figure 6. Gate tunnelling current density. Similar impact of the transition layer is observed, as in figure 6.

4. Conclusions

We elaborated on a simple procedure for obtaining a real-space-dependent band-gap and band-offsets from DFT calculations of the electronic structure of the oxide interface. Despite the inherent uncertainties associated with the translation of the microscopic data into continuum, macroscopic parameters, a more realistic picture of the band-gap transition is obtained, suitable for device modelling and simulation. Any of the simulated *realistic* barrier models exaggerates the magnitude of the impact that we have previously simulated with *linear* barrier model (up to 6 Å transition width). The spread in the characteristics, due to the uncertainty in the *realistic* profile, may be regarded as an uncertainty in the transition width t_{TR}, of a *linear* barrier model between 6 – 7 Å.

5. References

[1] Muller D A, Sorsch T, Moccio S, Baumann F H, Evans-Lutterodt K and Timp G 2000 *Nature* **399** 758; Hirose K, Nohira H, Azuma K and Hattori T 2007 *Prog Surf Sci* **82** 3
[2] Watanabe H, Matsushita D and Muraoka K 2006 *IEEE Trans Elec Dev* **53** 1323
[3] Stadele M, Sacconi F, Di Carlo A and Lugli P 2003 *J. Appl Phys* **93** 2681
[4] Markov S, Sushko P V, Roy S, Fiegna C, Sangiorgi E, Shluger A and Asenov A 2008 *Phys Stat Sol (A)* **205** 1290
[5] Yamasaki T, Kaneta C, Uchiyama T, Uda T and Terakura K 2001 *Phys Rev B* **63** 115314; Giustino F and Pasquarello A 2005 *Surf Sci* **586** 183; Xue K, Ho H P and Xu J B 2007 *J Phys D: Appl Phys* **40** 2886

Journal of Physics: Conference Series **242** (2010) 012011 doi:10.1088/1742-6596/242/1/012011

Electron velocity decline in Si nanoscales MOSFETs with the shortening of gate length

A. Islam, B. Benbakhti and K. Kalna

Device Modelling Group, Department of Electronics & Electrical Engineering, University of Glasgow, Glasgow G12 8LT, United Kingdom

E-mail: a.islam@elec.gla.ac.uk

Abstract. The ensemble Monte Carlo device simulations are employed to obtain I-V characteristics for the 25 nm gate length template Si MOSFETs designed by the SiNANO consortium. The simulated I_D-V_G characteristics are compared against previous results from various Monte Carlo device codes [Fiegna C et al., in Proc. SISPAD 2007, pp. 57-60 (Springer Vienna, 2007)]. After this verification, we have scaled the transistor laterally only from a gate length of 25 nm to gate lengths of 20, 15, 10 and 5 nm. We have then monitored the average electron velocity and sheet density along the channel at a supply voltage of 1.0 V in order to gain an insight into the degradation of the injection velocity experimentally observed in various Si transistor architectures when the gate length is scaled into deep sub-50 nm dimensions. We have found a substantial decrease of the overall velocity profile along the channel including a decrease of the peak velocity when the channel is smaller than 15 nm while the drive current is able to sustain its increase thanks to the increasing velocity at the drain side.

1. Introduction

The ITRS [1] predicts that a scaling of the bulk, planar Si CMOS technology will continue, and that in 2016, the gate length in multiple processor units (MPU) will reach 9 nm. The desire to continue the scaling of the planar technology is driven by lower costs when compared to non-planar concepts using multi-gate architectures [2]. The ultimate goal is to achieve near to ballistic transport with a continued gate length reduction [3]. However, experimental evidence suggests that carrier mobility and injection velocity will dramatically lower at very small gate lengths thus prohibiting the possibility of reaching the ballistic regime [2].

In this work, we monitor the average electron velocity along the channel in the 25 nm SiNANO n-channel MOSFETs [4] for decreasing the gate lengths from 25 nm to 20 nm, 15 nm, 10 nm, and 5 nm using ensemble Monte Carlo (MC) device simulation. The MC simulations include all relevant scattering mechanisms in Si [5], and includes self-consistent Fermi-Dirac statistics [6] and static screening model for the ionised impurity scattering [7], interface roughness scattering and quantum correction using the effective quantum potential method [6]. The electron velocity has been monitored at a working on-current condition of the transistor for the supply voltage of 1.0 V which means that we have applied a drain voltage of 1.0 V and gate voltage of 1.2 V presuming a required threshold voltage of 0.2 V. Figure 1 shows the cross-section of a 25 nm gate length n-channel SiNANO MOSFET device with an Oxynitride (ON) gate stack and a metal gate which was developed as a template transistor for the SiNANO consortium [8] in order to compare various simulation approaches and tools. It has source and drain ohmic contacts on

© 2010 IOP Publishing Ltd

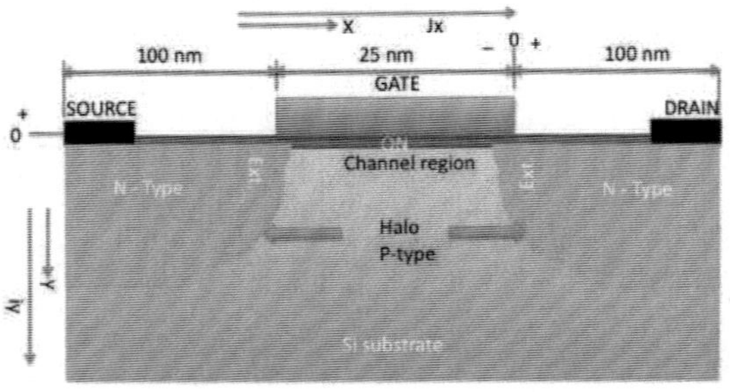

Figure 1. Cross-section of the investigated 25 nm gate length n-channel SiNANO MOSFETs, whose sides are parallel to the x- and y-axes and was a starting point for the gate shortening study.

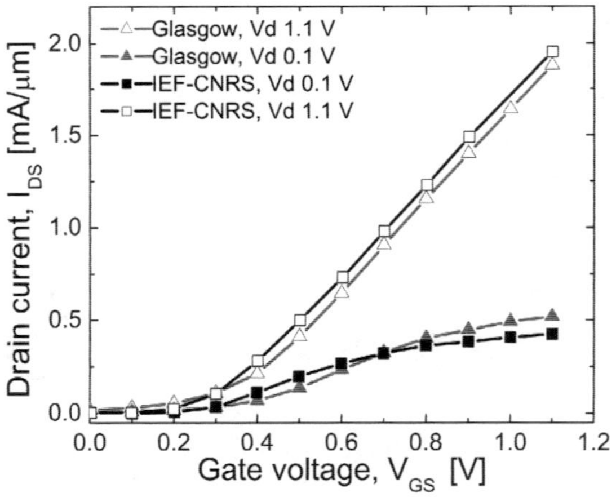

Figure 2. $I_D - V_G$ for the 25 nm gate length Si MOSFET biased at $V_D = 0.1$ V and $V_D = 1.1$ V, comparing our results with those from [4].

top of the highly doped $n+$ regions with a concentration of 1×10^{20} cm^{-3} and extension regions are doped to 1×10^{20} cm^{-3}, halo p-type doping with a concentration of 8×10^{18} cm^{-3}, and a Si substrate with a p-type doping concentration of 3×10^{18} cm^{-3}. Because of the overlap of extension doping with the gate region at the source and drain side of the transistor, the channel region length (which is also called the effective channel length) is slightly shorter than physical gate length. The transistor has a Si/ON interface which is assumed to be perpendicular to the y-axis. The ON gate oxide has a thickness of 1.6 nm and dielectric constant of $\epsilon_{OX} = 7$.

2. I_D-V_G characteristics

Figure 2 compares MC simulated $I_D - V_G$ characteristics for the 25 nm gate length Si MOSFET at low and high drain biases of 0.1 V and 1.1 V. The MC simulations include Fermi-Dirac statistics by self-consistently calculating Fermi level and electron temperature at each MC time step. The obtained Fermi level and electron temperature is then used in a static screening model in the ionised impurity scattering. The MC device simulations also include interface roughness scattering based on Ando's model [6]. The obtained $I_D - V_G$ characteristics are also compared to those obtained from MC simulations performed with IEF-CNRS MC code [4]. The full and open squares are the results from the IEF-CNRS code, and the full and open triangles are the results from our simulations (Glasgow), in the both cases biased at low and high drain voltages of 0.1 V and 1.1 V, respectively, showing a very good agreement.

Figure 3 shows the average electron velocity and carrier density along the channel of the Si MOSFET with a gate length of 25 nm when considering Fermi-Dirac statistics using the Ridley's model [7] and the static screening model for the ionised impurity scattering [6]. The average electron velocity has been monitored at a drain voltage of 1.1 V and a gate voltage of 1.0 V. The full squares in Fig. 3 shows that the average electron velocity steeply increases at the end of the source region (the beginning of the gate) and then gradually increases along the channel. This increase is caused by the acceleration of electrons at very high, fringing electric field present at the beginning of the gate. The electron velocity will reach its peak of 2.3×10^7 cm/s in the channel, close to the drain side, when electrons gain high energies and start to suffer from

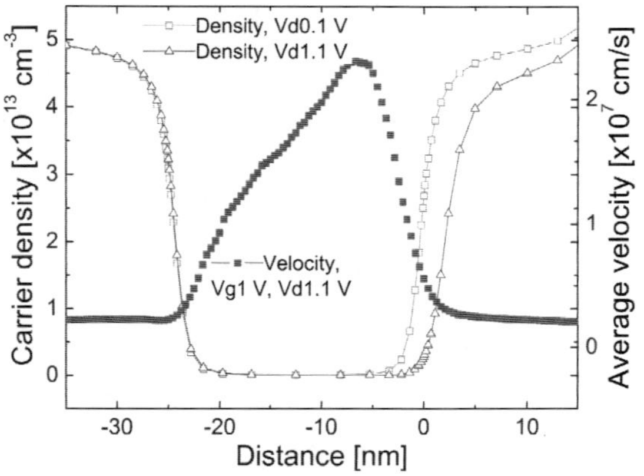

Figure 3. Average velocity and electron density along the channel of the 25 nm Si MOSFET. The beginning of the gate is at −25 nm and the end is set at zero.

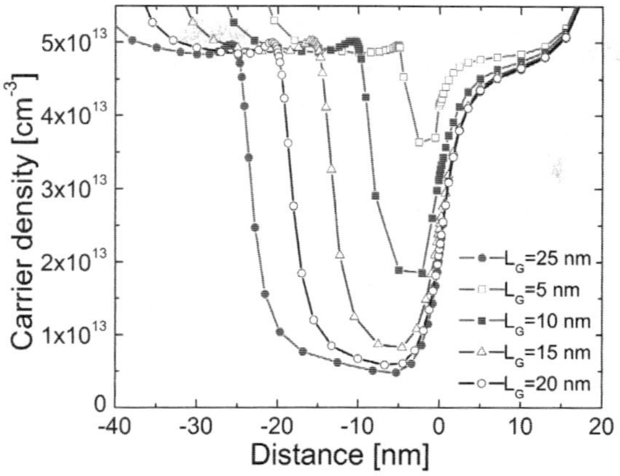

Figure 4. Carrier density as a function of the distance along the channel biased at V_G-V_T = 1.0 V, V_G = 1.0 V for the 25nm Si MOSFET.

the increased scattering with acoustic phonons and from the start of the scattering with optical phonons. The reason for the subsequent a sharp decline in the electron velocity is this start of the electron scattering with inter-valley f- and p-type non-polar optical phonons which affects the electrons with large energies.

3. Carrier density and average velocity

After this verification of our MC device code, we have performed simulations in order to analyse the performance of scaled SiNANO MOSFETs. In these simulations, we have also employed quantum corrections using the effective quantum potential method [9]. The smoothing parameter required for the effective quantum potential has been calibrated against 1D Poisson-Schrödinger simulations [6]. Figure 4 shows the carrier density as a function of the distance along the channel biased at a gate voltage of 1.2 V, and a drain voltage of 1.0 V for different gate lengths. The electron density is below $1 \times 10^{13}\text{cm}^{-3}$ along the 25 nm length channel while, after the reducing the channel length and thus reducing the depletion region, the density in the channel gradually increases finally reaching $3.5 \times 10^{13}\text{cm}^{-3}$ at a reduced gate length of 5 nm. The average velocity profiles obtained when the gate length of the devices is reduced from 25 nm to 20 nm, 15 nm, 10 nm, and 5 nm are reported in the Fig. 5. The peak velocity reaches 2.1×10^7 cm/s in the transistor with a gate length of 25 nm, and it will increase during the scaling to its maximum of about 2.2×10^7 cm/s when the gate length is 15 nm. After this, the peak velocity starts to decline to 1.48×10^7 cm/s for the gate length of 5 nm. The results reveal that the non-equilibrium electron transport in the devices with the gate length less than 15 nm is so intense that it causes a very quick increase of the electron energy which consequently starts the electron interactions with inter-valley f- and p-type non-polar optical phonons. This process is assisted by the re-distribution of electron momenta through the long-range Coulomb interaction because the heavily doped source and drain are very close [11]. This then leads to a quick loss of electrons energy and decline in their velocity. The inset of the Fig. 5 shows the peak velocities versus gate length showing a maximum of 2.2×10^7 cm/s for the 15 nm gate length, while it declines to 1.3×10^7 cm/s for the 5 nm gate length device. Figure 6 shows that the drain current obtained in the shortening of gate lengths at $V_G = 1.2$ V and of $V_D = 1.0$ V for scaled bulk Si MOSFETs. The drain current gradually increases with decreasing the gate lengths of the device even when the transistor gate length is scaled below 15 nm. The fact that the drain current is able to

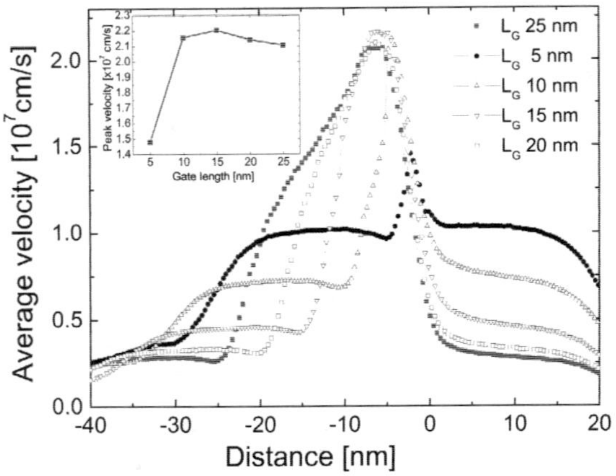

Figure 5. Electron velocities along the channel for indicated gate lengths at V_G-V_T = 1.0 V and V_G = 1.0 V. Zero distance is set at the end of the gate. The inset shows the peak velocities vs. the gate length.

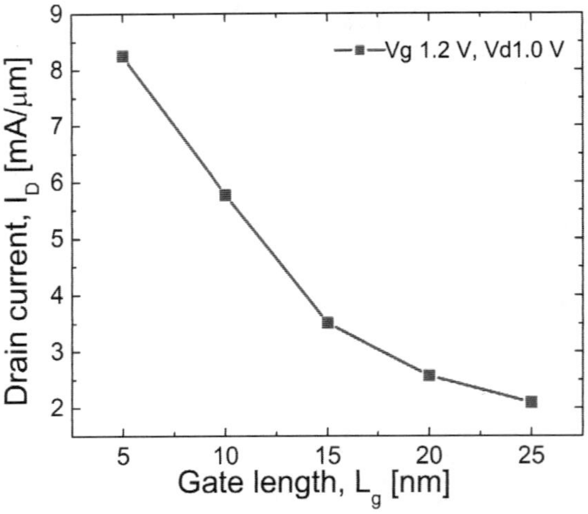

Figure 6. Drain current as a function of the gate length at a supply voltage given by V_G-V_T = 1.0 V and V_G = 1.0 V for the scaled bulk SiNANO MOSFETs.

sustain its increase is thanks to the increasing drain velocity at the drain side of the device. This demonstrates that the ballistic theory [10] which predicts that the drain current is govern by the injection velocity at the source fails at nanoscale transistor dimensions [11].

4. Conclusions

Monte Carlo (MC) device simulations are used to analyse the performance of scaled 25 nm gate length SiNANO *n*-channel MOSFETs. We studied the physical mechanisms which determine the average electron velocity and sheet density as a function of the horizontal position in the channel of MOSFETs with decreasing the gate length. We have found that the drive current is steadily increasing in the scaling process despite a decline in the peak velocity from 15 nm gate length as well as the occurrence of a velocity saturation region at the source side of the channel. The decrease of the average velocity including drop in the peak velocity is caused by enhanced electron scattering with non-polar optical phonons as they reach quickly very high energies. This process of electron energy loss is also strongly assisted by the long-range Coulomb interaction from the heavily doped source/drain regions [11]. The drain current increase is entirely thanks to the steadily increasing velocity at the drain side of the scaled transistors.

5. References

[1] International Technology Roadmap for Semiconductor 2007, [http://public.itrs.net]
[2] Fleury D, Bidal G, Cros A, Boeuf F, Skotnicki T, Ghibaudo G 2009, VLSI Symp. Tech. Dig. Pap., pp. 16-17.
[3] Yu B, Wang H, Joshi A, Xiang Q, Ibok E, and Lin M 2001, IEDM Tech. Dig., pp. 937-40.
[4] Fiegna C, et al. 2007, in Proc. SISPAD 2007, pp. 57-60 (Springer Vienna).
[5] Jacoboni C, and Lugli P, *The Monte Carlo Method for Device Simulation*, ISBN 3-211-82110-4, (1989).
[6] Kalna K, L. Yang and A. Asenov 2005, in Proc. ESSDERC 2005 (Grenoble, France) 169-172.
[7] El-Ghanem H M A and Ridley B K, J. Phys. C: Solid State Phys. **13**, 2041 (1980).
[8] SiNANO Consortium [http://www.sinano.org/].
[9] Ferry DK, Akis R, and Vasileska D, IEDM Tech. Dig., San Francisco, Dec 10-13, pp. 287-290 (2000).
[10] Lundstrom M 1997, IEEE Electron Device Lett., **18**, 361-363.
[11] Fischetti MV, et al. 2009, J. Comput. Electron. **8**, 60-77.

Journal of Physics: Conference Series **242** (2010) 012012 doi:10.1088/1742-6596/242/1/012012

Theoretical modelling of InGaAs quantum rods: terahertz intraband absorption and its dependence on rod height

N. Prodanović[1,2], N. Vukmirović[3], D. Indjin[2], Z. Ikonić[2], and P. Harrison[2]

[1]Faculty of Electrical Engineering, University of Belgrade, Belgrade, Serbia
[2]School of Electronic and Electrical Engineering, University of Leeds, Leeds, UK
[3]Computational Research Division, Lawrence Berkeley National Laboratory, Berkeley, USA

Abstract. The electronic structure and the intraband optical properties of self-assembled quantum rods were analysed using a theoretical model based on strain-dependent eight-band $\mathbf{k} \cdot \mathbf{p}$ method. It was found that the intraband optical absorption spectrum for the radiation polarised in the growth direction has one strong peak caused by the transition from the ground state to the state confined in the rod with one node in the growth direction. On the other hand the absorption of in-plane polarised radiation is weaker and is determined by the transitions to mixed quantum well - quantum rod states. The most prominent transitions fall into the technologically relevant terahertz region of the electromagnetic spectrum and can be tailored by engineering the quantum rod height.

1. Introduction

Self-assembled semiconductor quantum dots have attracted a lot of interest in the past decades due to their applications in electronic and optoelectronic devices, which are expected to exhibit better performance than similar devices based on quantum wells. Typical self-assembled quantum dots grown in Stranski-Krastanov mode have height-to-diameter aspect ratio significantly smaller than one. Novel quantum nanostructures, so called quantum rods or quantum posts [1, 2, 3, 4] have recently been realised. Such structures were obtained by the alternating deposition of very short (of the order of monolayer) layers of InAs and GaAs. Subsequent intermixing of InAs and GaAs leads to the InGaAs quantum rod with large height-to-diameter ratio. The height of these structures, and consequently their aspect ratio, can be tailored by the number of deposited layers in contrast to conventional quantum dots where it is more difficult to control the geometry. It is therefore of great importance to develop a theoretical description of electronic and optical properties which would enable the optimisation of quantum rod structures. Initial work in this area was focused on the calculation of energy levels and interband optical properties [1, 2, 5]. In this work, we focus on intraband optical properties, which are of particular interest since the transitions are in the technologically relevant terahertz region and can be adjusted by the height of the quantum rod.

© 2010 IOP Publishing Ltd

Journal of Physics: Conference Series **242** (2010) 012012 doi:10.1088/1742-6596/242/1/012012

2. Theoretical Considerations

The electronic structure of quantum rods was modelled using the eight-band envelope function $\mathbf{k} \cdot \mathbf{p}$ model [6, 7, 8, 9, 10]. Strain distribution was taken into account using the continuum mechanical model [11]. Numerical calculation of the components of the strain tensor was performed using the finite element method [12]. Rods of cylindrical shape were considered and the axial approximation [13] was introduced. In such a way, the problem was reduced from a three dimensional one to a two dimensional one. The Hamiltonian eigenvalue problem was then solved using the orthonormal function expansion method where the basis was formed from the direct product of Bessel functions in the radial direction and the plane waves in the growth (z) direction. The cylindrical symmetry of the rods introduces a good quantum number of the $z-$component of the total quasi-angular momentum m_f which takes the half-integer values [13, 14]. The optical transition selection rules then allow only the transitions with $\Delta m_f = 0$ in the case of $z-$polarised radiation and $|\Delta m_f| = 1$ for $x-$polarised radiation.

The structures at low temperatures were considered here and therefore it was assumed that only the ground state in the conduction band is significantly occupied. The intraband optical absorption spectrum was therefore calculated by adding the contributions from the transitions from the ground state to each of the excited states. The inhomogeneity of the quantum rod ensemble was taken into account by assuming a Gaussian lineshape with a standard deviation equal to 10% of the transition energy for each of the transitions.

3. Numerical Results and Discussion

The quantum post geometrical parameters and material composition were taken in accordance with experiments from the literature. The diameter was taken to be equal to 10 nm [1]. The height of the rod was varied from 20 nm to 40 nm, which is the typical range of quantum rod heights [1, 2], with a step of 5 nm. The InAs/GaAs short-period superlattice, away from the quantum rod, becomes an InGaAs quantum well layer with an In composition of 16% [1] during the growth of the structure. The In content in the quantum rod is approximately 45% [1]. Such a structure is schematically shown in the bottom part of Fig. 1.

The energy level diagram and the wavefunctions of the most relevant states for the transitions from the ground state are shown in top part of Fig. 1 for the 30 nm high rod. The absorption of $z-$polarised radiation from the ground state, which has no nodes, is strongest to a state which has one node in the $z-$direction. This state arises due to quantum confinement effect in the $z-$direction and its energy is therefore directly determined by the rod height. As a consequence, the intraband optical absorption spectrum of the $z-$polarised radiation can be directly tailored by the quantum rod height, as can be seen from Fig. 2. For the largest quantum rods (40nm), when the height is significantly larger than the diameter, this state is the first excited state. Therefore, the absorption spectrum for $z-$polarised radiation is then fully determined by the transition from the ground to first excited state. This situation is dual to the situation in conventional flat-shaped quantum dots where the absorption spectrum of $x-$ polarised radiation is dominated by the transition to the pair of first excited states [13]. In these dots the first excited state arises due to confinement in the lateral direction and therefore it is $x-$ rather than $z-$polarised radiation that can excite the electron from the ground state.

On the other hand, when the absorption of $x-$polarised radiation in quantum rods is concerned, one would normally expect (in a quantum rod with infinite potential barrier) that it takes place to a state with one node in the lateral direction. In quantum rods that are narrow such a state should have an energy which is quite high and one would therefore expect that absorption of $x-$polarised radiation takes place at much larger photon energies than in the case of $z-$polarised radiation. However, this is not the case in the structure considered here, as can be seen from Fig. 2. One should note that the InGaAs quantum rod confinement potential is rather shallow, as can be seen from Fig. 1. Furthermore, the InGaAs quantum rod is embedded in an

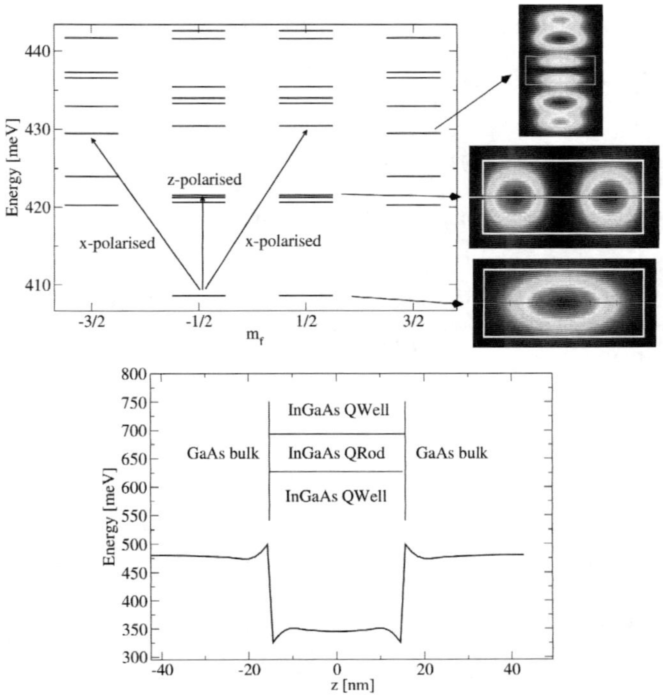

Figure 1. Top: Energy level diagram of the states with $|m_f| \leq 3/2$ for a 30 nm high quantum rod. The transitions that mostly determine the intraband absorption spectrum for incident $x-$ and $z-$polarised radiation are marked. The contour plots of the wavefunctions of the states involved in these transitions are shown on the right. White lines represent the quantum rod boundary and its axis. Bottom: Effective on-axis potential profile and the schematic view of the structure.

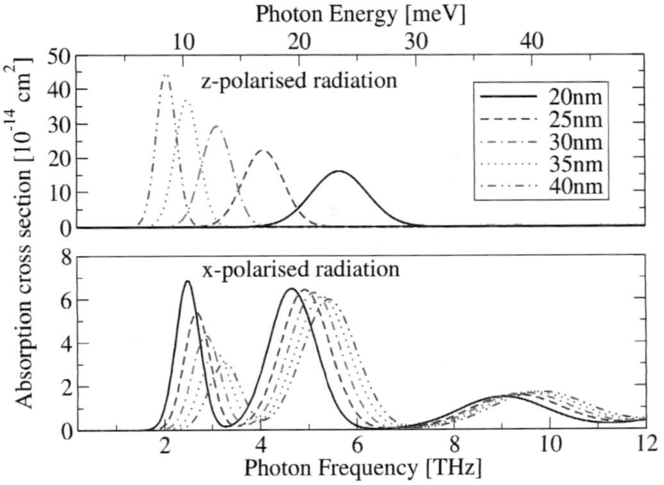

Figure 2. Intraband optical absorption cross section of quantum rods in the case of $z-$ polarised radiation (top) and $x-$polarised radiation (bottom). The height of each of the quantum rods is indicated in legend.

InGaAs quantum well, which makes the lateral confinement even shallower. As a consequence, the rod with such a shallow potential cannot accommodate the state where the wavefunction has a node in the radial direction and is in the same time confined within the rod. In the rods considered here, the absorption of $x-$polarised radiation predominantly takes place to the states, such as the one illustrated in Fig. 1 in the case of the 30 nm rod. As can be seen from Fig. 1 the wavefunction of this state is not confined at all to the interior of the rod. This is therefore a mixed quantum rod - quantum well state, with no nodes in the $z-$direction. The position of such states with respect to the ground state is therefore much less affected by the quantum rod height than for the states with nodes in the $z-$direction. As a consequence, the shifts in the intraband optical absorption spectrum of $x-$polarised radiation are much smaller than in the case of $z-$polarised radiation. Furthermore, the absorption spectrum for $x-$polarised radiation is much weaker than for $z-$polarised radiation, which is fully in-line with the delocalised nature of the excited state to which the absorption takes place.

4. Conclusion

To conclude, we have used the eight band $\mathbf{k}\cdot\mathbf{p}$ model to calculate the electronic and optical properties of a new class of semiconductor quantum nanostructures: self-assembled quantum rods. We found that the positions of the absorption peaks can be tailored by the choice of rod height. The rods respond quite differently to $x-$ and $z-$polarised radiation. The most prominent transitions fall in the (2–6) THz spectral region which is the region where existing terahertz quantum cascade lasers operate. Consequently, the possibility is opened for integration of quantum rod terahertz detectors with quantum cascade lasers for terahertz sensing applications.

Acknowledgments

N.P. would like to thank the School of Electronic and Electrical Engineering, University of Leeds, Research Internship Scheme for financial support. D.I. acknowledges the support from NATO Science for Peace Collaborative Linkage Grant Programme.

References

[1] Li L, Patriarche G, Chauvin N, Ridha P, Rossetti M, Andrzejewski J, Sek G, Misiewicz J and Fiore A 2008 *IEEE J. Select. Topics in Quantum Electron.* **14** 1204–1213
[2] He J, Krenner H J, Pryor C, Zhang J P, Wu Y, Allen D G, Morris C M, Sherwin M S and Petroff P M 2007 *Nano Lett.* **7** 802–806
[3] Motyka M, Sek G, Ryczko K, Andrzejewski J, Misiewicz J, Li L H, Fiore A and Patriarche G 2007 *Appl. Phys. Lett.* **90** 181933
[4] Ridha P, Li L, Fiore A, Patriarche G, Mexis M and Smowton P M 2007 *Appl. Phys. Lett.* **91** 191123
[5] Saito T, Ebe H, Arakawa Y, Kakitsuka T and Sugawara M 2008 *Phys. Rev. B* **77** 195318
[6] Bahder T B 1990 *Phys. Rev. B* **41** 11992–12001
[7] Tomić S, Sunderland A G and Bush I J 2006 *J. Mater. Chem.* **16** 1963–1972
[8] Vukmirović N, Indjin D, Jovanović V D, Ikonić Z and Harrison P 2005 *Phys. Rev. B* **72** 075356
[9] Mlinar V, Tadić M, Partoens B and Peeters F M 2005 *Phys. Rev. B* **71** 205305
[10] Pryor C 1998 *Phys. Rev. Lett.* **80** 3579–3581
[11] Stier O, Grundmann M and Bimberg D 1999 *Phys. Rev. B* **59** 5688–5701
[12] Davies A J 1980 *The Finite Element Method* (Clarendon Press, Oxford)
[13] Vukmirović N, Gačević Ž, Ikonić Z, Indjin D, Harrison P and Milanović V 2006 *Semicond. Sci. Technol.* **21** 1098–1104
[14] Vukmirović N, Indjin D, Ikonić Z and Harrison P 2006 *Appl. Phys. Lett.* **88** 251107

Modeling of intraband absorption for quantum dot-in-well structures with low computational cost

B H Hong[1], S I Rybchenko[1], I E Itskevich[1], S K Haywood[1], C H Tan[2], P Vines[2], and M Hugues[2]

[1]Department of Engineering, University of Hull, Cottingham Road, Hull, HU6 7RX, UK
[2]Department of Electronic and Electrical Engineering, University of Sheffield, Mappin Street, Sheffield, S1 3JD, UK

E-mail: b.h.hong@eng.hull.ac.uk

Abstract. Much effort has been committed to development of quantum-dot-based infrared photodetectors owing to their potential for normal-incidence absorption and low dark current. Quantum-dot-in-well structures offer additional advantages, such as better wavelength tunability and improved carrier collection. This system presents a challenge for modeling of electronic structure, as it requires solution for a complex system (quantum dot plus quantum well) with both discrete levels and the continuum energy spectrum. The Green's function method, mostly used for such problems, has very high computational cost. Here we use the Finite Element Method to model intraband absorption spectra of quantum-dot-in-well structures within the effective mass approximation.

1. Introduction

Quantum dot infrared photodetectors (QDIPs) have attracted much interest in recent years [1-4], aiming to overcome the limits of HgCdTe and quantum well (QW) detectors. Due to the three-dimensional carrier confinement in the quantum dot (QD), these devices have potential advantages, such as sensitivity to normal incidence radiation and low dark current. Quantum-dot-in-well (DWELL) structures offer additional advantages, such as better control over operating wavelength and improved carrier collection. The DWELL devices rely on transitions between the bound states in the dot and the subbands in the surrounding QWs. The QD-based intersubband photodetector is a promising technology for the 3-12 μm wavelength region.

This system presents a challenge for modeling of the electronic structure, as it combines zero-dimensional QDs and 2-dimensional QWs with discrete and continuous energy spectra, respectively. The Green's function method is the tool mostly used for such problems [5-6], but it has a very high computational cost. Here we use a simplified approach employing the Finite Element Method within the effective-mass approximation. We find that it has both low computational cost and good accuracy, thus allowing modeling of the intraband absorption spectra for a wide range of QD parameters.

2. Experimental details

We investigated a DWELL system, in which InAs dots are embedded asymmetrically in a GaAs well, surrounded by a thick $Al_{0.1}Ga_{0.9}As$ barrier. A schematic diagram of the structure is shown in figure 1.

© 2010 IOP Publishing Ltd

Journal of Physics: Conference Series **242** (2010) 012013 doi:10.1088/1742-6596/242/1/012013

Result of the transmission electron microscopy (TEM) on the sample is shown in figure 2. The dots were found to have a truncated shape with height of 7 nm and width of 35 nm on average. The dot density was estimated as 3 x 10^{10} cm^{-2}, corresponding to 60 nm of lateral separation between the dots.

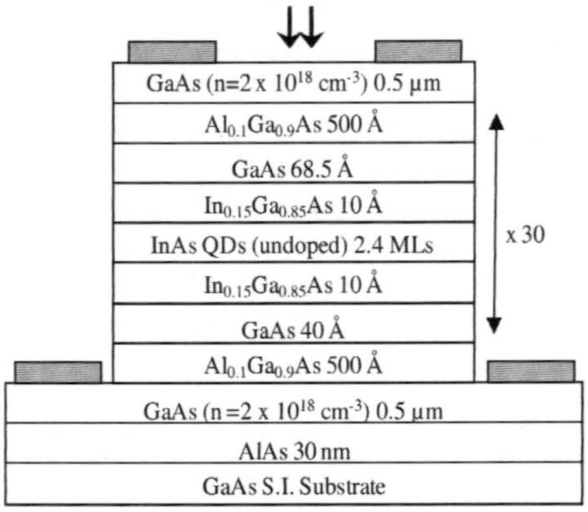

Figure 1. Schematic diagram of the DWELL structure.

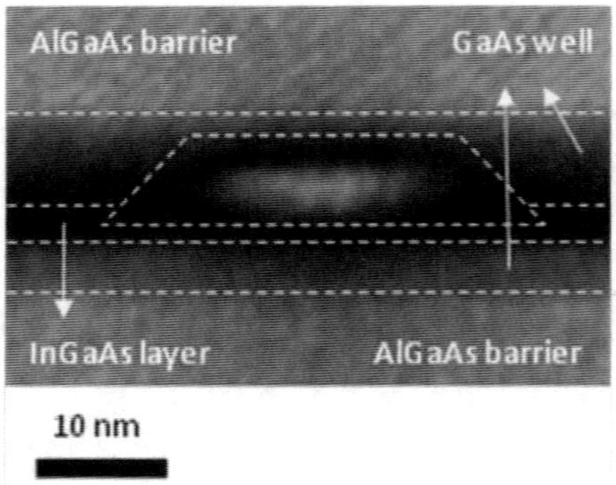

Figure 2. TEM image of the sample. Dashed lines outline the structure geometry.

3. Modeling

For all calculations, we employed the Finite Element Method using COMSOL Multiphysics commercial software. Strain effect was taken into account in the continuum-elasticity approximation using standard deformation potential theory. All material parameters were taken from reference [7].

Since the inter-dot distance is more than 50 nm both laterally and vertically, the interaction between the dots is negligible. Therefore, modeling for an isolated QD is a good approximation. We approximated the QW by a large flat quantum box. This allowed us to calculate the quasi-continuum energy spectrum and relevant wave function by solving directly the 3D effective mass problem.

The model includes layers of square boxes with different compositions as described in figure 1. We assumed a dot with truncated-cone shape, positioned in the middle of the thin $In_{0.15}Ga_{0.85}As$ strain-reducing layer. The wetting layer is considered negligible, as it actually coincide with the strain-reducing layer. As illustrated in figure 2, the dot is positioned asymmetrically in the GaAs well.

Because of In diffusion, the $In_{1-x}Ga_xAs$ dot composition is expected instead of pure InAs. The dot composition was estimated prior to modeling of the intraband absorption spectra, assuming homogeneous composition in the dot. The transition energies between electron and hole levels were modeled for isolated dot using the conventional 8-band strain-dependent **k·p** Hamiltonian [8-9] and compared to the experimental photoluminescence (PL) data.

For the entire DWELL structure, we modeled the quasi-continuum spectrum of the subbands using only the one-band strain-dependent effective mass approximation. The energies and matrix elements of the transitions between the ground state (GS) of the dot and the excited states of the DWELL structure were determined. (Note that only the GS of the dot is normally occupied in such a device.) Finally, the intraband absorption spectra were simulated by superimposing the Gaussian broadening on the calculated transitions, to account for inhomogeneous distribution in the dot size and geometry.

4. Results and Discussions

The PL data in figure 3 show two distinct peaks at 0.99 eV and 1.04 eV. The low energy peak (PL 1) corresponds to the transition between the ground states of electrons and holes (D0-V0), while the higher energy peak (PL 2) is due to transition between their excited states (D1-V1). When compared to the PL data, the results of multi-band **k·p** model for isolated $In_{1-x}Ga_xAs$ QD in the GaAs matrix show that the percentage of Ga atoms in the dot is approximately 25-30% (figure 4).

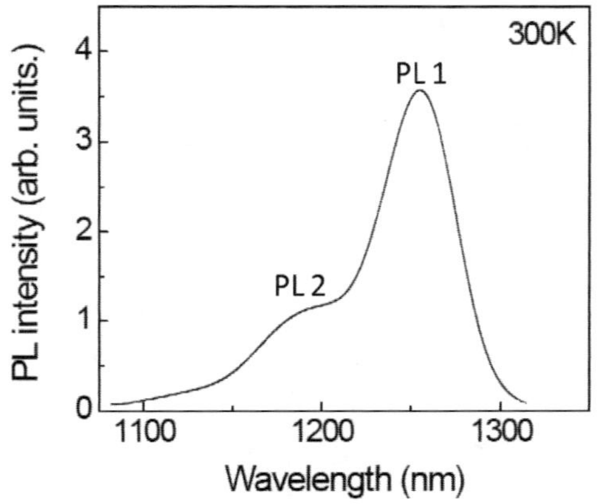

Figure 3. Experimental PL spectra at 300 K.

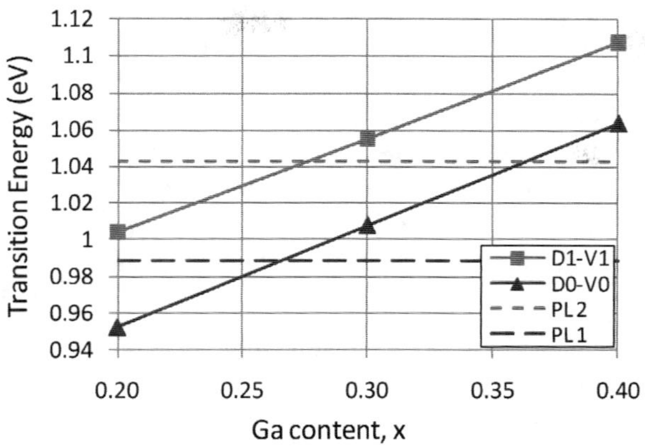

Figure 4. Calculated transition energies (D1-V1 and D0-V0) for the $In_{1-x}Ga_xAs$ QD as a function of composition at 300 K. Dashed lines show energies of the experimental PL transitions.

We have modeled the optical spectra of the DWELL structures for a wide range of QD geometrical parameters. We found that the best agreement with experiment is achieved for the QD parameters very close to that obtained from the TEM data (figure 2). The modeled spectra for a DWELL with $In_{0.75}Ga_{0.25}As$ QD using the experimentally deduced parameters are shown in figure 5. The vertical lines are individual transitions while the bold lines show the absorption spectra with superimposed Gaussian broadening of 5 meV. The results show expected strong absorption for *z*-polarized light as compared to the in-plane polarization. The normal-incidence experiments correspond to in-plane polarized light, figure 5(a). From modeling, the 1st and 2nd subbands are present in the well, with transitions to them in the ranges of 4.5-5.9 μm and 3.9-4.4 μm, respectively. Peaks below 3.9 μm correspond to bound-to-continuum transitions. Also from calculations, there are strong absorption peaks corresponding to the transitions between the ground and excited dot states (GS-ES), with peaks at around 6.7 μm, 7.3 μm and 8.5 μm.

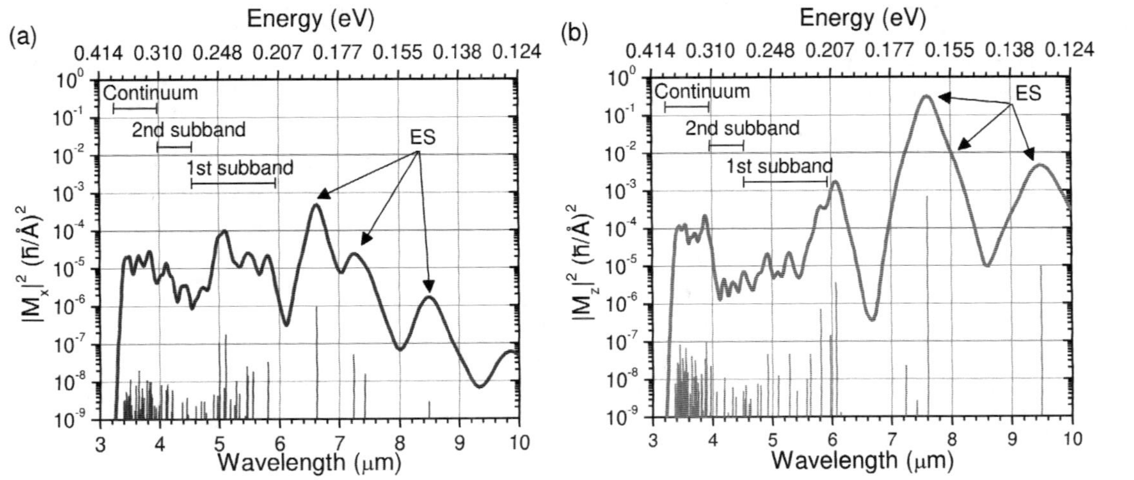

Figure 5. Calculated zero-bias transitions at 77 K for polarizations along (a) *x*/*y*, and (b) *z* directions, respectively. Bold lines simulate absorption spectra with superimposed broadening of 5 meV.

Figure 6 shows the experimental photocurrent spectra at various bias voltages. Note that distortions around 4.3 μm and 5-7 μm are due to atmospheric absorption. Peak positions are in good agreement with modeling. At low positive bias, figure 6(a), peaks are observed at 3.5 μm (GS to continuum), 4.2 μm (GS to 2nd subband), and 5 μm (GS to 1st subband). At higher positive bias, peaks at around 6 μm and 7.5 μm emerge, which we associated with the GS-ES transitions. These transitions are also observed under negative bias, figure 6(b). The subband-related transitions are shifted towards longer wavelength in this case, due to applied electric field and asymmetric dot position in the well.

The photocurrent measurements were performed under normal incidence radiation. However, due to scattering, contribution from *z*-polarized component is likely. This is particularly important for the GS-ES transitions (e.g. at 6 μm), for which the *z*-polarized absorption is at least three orders of magnitude stronger than in-plane polarized one.

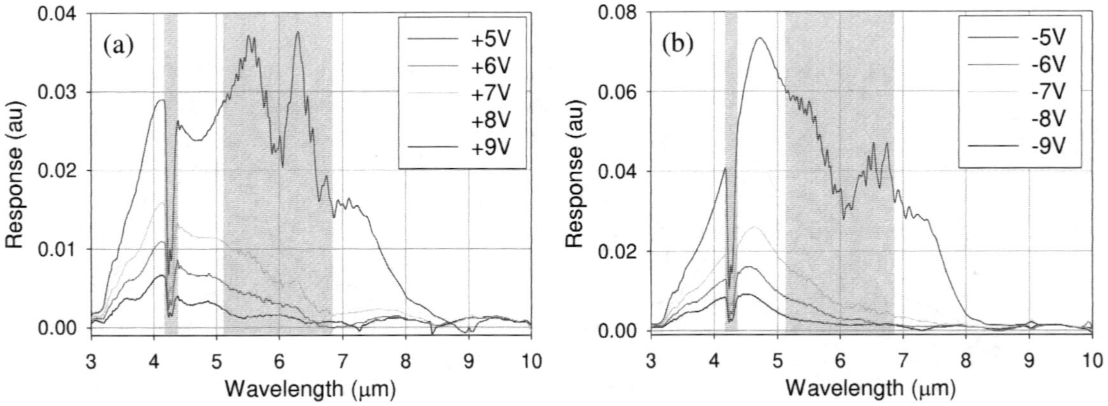

Figure 6. Photocurrent spectra obtained at 77 K under (a) positive bias, and (b) negative bias. Shaded bands at 4.3 μm and 5-7 μm show atmospheric absorption band.

5. Summary

Using a model with low computational cost, we explored electronic spectra of a DWELL structure for a wide range of QD geometrical parameters. Our results show good agreement with the experimental data. The model provides an efficient tool for design of the DWELL systems.

References

[1] Finkman E, Maimon S, Immer V, Bahir G, Schacham S E, Fossard F, Julien F H, Brault J and Gendry M 2001 *Phys. Rev. B* **63**, 045323
[2] Vukmirović N, Ikonić Z, Savić I, Indjin D and Harrison P 2006 *J. Appl. Phys.* **100**, 074502
[3] Lim H, Zhang W, Tsao S, Sills T, Szafraniec J, Mi K, Movaghar B and Razeghi M 2005 *Phys. Rev. B* **72** 085332
[4] Krishna S 2005 *J. Phys. D: Appl. Phys.* **38** 2142
[5] Naser M A, Deen M J and Thompson D A 2007 *J. Appl. Phys.* **102** 083108
[6] Putz V, Ede C and Scrinzi A 2007 *Phys. Rev. B* **76** 035318
[7] Vurgaftman I, Meyer J R and Ram-Mohan L R 2001 *J. Appl. Phys.* **89** 5815
[8] Bahder T B 1992 *Phys. Rev. B* **45** 1629
[9] Pryor C 1998 *Phys. Rev. B* **57** 7190

Journal of Physics: Conference Series **242** (2010) 012014 doi:10.1088/1742-6596/242/1/012014

Controlling dopant solubility in semiconductor alloys

A. Höglund,[1] C.W.M Castleton,[2] O. Eriksson[1] and S. Mirbt[1]

[1]Condensed Matter Theory Group, Department of Physics, Uppsala University, Box 530, SE-751 21 Uppsala, Sweden

[1]School of Science and Technology, Nottingham Trent University, Nottingham, NG11 8NS, UK

E-mail: Christopher.Castleton@ntu.ac.uk

Abstract. We consider the formation energies and stabilities of dopants in semiconductor alloys. We show that they are not bounded by the formation energies in the related pure materials. On the contrary, by tuning the alloy composition, dopant solubility can be increased significantly above that in the pure materials. Furthermore, it is not always necessary to carry out full defect calculations in alloy supercells, since good estimates of the formation energies at the most stable substitution sites can be obtained by calculating the formation energies in the various component *pure* materials, but *strained* to the lattice parameter of the alloy.

1. Introduction

Atomic scale point defects play a defining role in a great many industrially and environmentally relevant materials properties and processes, from electronic doping in semiconductors, to hydrogen transport within fuel cells, to the active sites in metal/metal oxide catalysts. Our ability to study such defect properties and processes from as many angles as possible is therefore extremely important. One very important angle is now the use of *ab initio* Density Functional Theory (DFT). [1] The key property usually calculated is the defect formation energy,

$$E_q^F = E_{def}^{Tot}(q) - E_{ideal}^{Tot} - \sum_i n_i \mu_i + q(\varepsilon_F + \varepsilon_V), \qquad (1)$$

where $E_{def}^{Tot}(q)$ and E_{ideal}^{Tot} are the total energies of a particular supercell or atomic cluster with and without the charge q defect. n_i atoms of type i and chemical potential μ_i are added while forming the defect. ε_F is the Fermi level, and ε_V is the valence band edge (VBE) measured relative to it. Almost all defect properties can be obtained from E_q^F, and the ways in which it varies with structure, lattice site, host composition, pressure, etc.

The calculation of E_q^F in the bulk of pure materials has become routine (see review [3]) although there remain problems due to the approximations required (see discussion in [4]). Calculations become more complex and expensive if we consider defects at surfaces (e.g. [5]) or interfaces between pure materials (rarely treated [3]), but overall a broad understanding of defects in pure materials has been achieved. However, many of the most important semiconductor materials are not pure at all, but alloys: either binary (e.g. $Si_{1-x}C_{1-x}$), tertiary (e.g. $Al_xGa_{1-x}As$), or quarternary (e.g. $In_xGa_{1-x}As_yP_{1-y}$). Calculations here are far more challenging. The simplest approach is to use the Virtual Crystal Approximation (VCA) [6], in which we replace, say, both the Al and Ga in $Al_xGa_{1-x}As$ with a pseudo-atom whose properties are a weighted average of those of As and Ga. The underlying assumption is that

© 2010 IOP Publishing Ltd

the properties of the alloy are a linear, or near linear interpolation between the properties of the pure materials, and local details do not matter. For the lattice parameter this is a rather good approximation, known as Vegard's Law [2]: the lattice constant a_0^{AB} of an alloy A_xB_{1-x} is given by $a_0^{AB} = xa_0^A + (1-x)a_0^B$. For some materials the VBE and conduction band edge (CBE) and hence the band gap also follow a similar relation, although more generally there is some curvature. The VCA would predict that in alloys the defect formation energies, E_q^F, vary monotonically, (probably linearly) between those of the pure materials.

However, in most cases [3] the details of local relaxations and the identity of nearby ions make a significant contribution to E_q^F. In a random alloy such as $Ga_xIn_{1-x}P$ some local environments will resemble InP and some GaP, with others in between. One might therefore anticipate a spread of E_q^F values, bounded by those in pure InP and GaP. However, in a recent study of dopants in both alloys and thin multilayers [7] we found that while there *is* a wide spread of E_q^F values, it is *not* bounded by those in the pure materials. Indeed, dopants can be far more stable in the alloy or multilayer than in either pure material, causing significant increases in solubility. Here, we report further results and details regarding the behaviour of dopants in disordered and ordered alloys. Since strain effects turn out to dominate, we consider the examples of Zn_{III}^- and Cd_{III}^- in $Ga_xIn_{1-x}P$. The cationic radii follow the order Zn ≈ Ga < In ≈ Cd, so we can study the substitution of both In and Ga with smaller, similar and larger radius dopants.

2. Computational details

We use plane wave *ab initio* DFT [1] within the Local Density Approximation (LDA) [10] together with ultrasoft pseudopotentials [11, 12] using the VASP code.[13] The indium $4d$ electrons are treated as core, and for charged defects a uniform compensating background maintains the charge neutrality of the supercell.[14]. All ions are relaxed fully. Where both ions and volume are relaxed this is done iteratively, alternating between relaxing the volume with fixed ions and relaxing the ions at fixed volume, until both are simultaneously converged. LDA lattice constants of 5.39 Å and 5.83 Å are used for GaP and InP [8, 9].

Calculations in strained pure materials use 216 atom cells. For the ordered phase of $Ga_{0.5}In_{0.5}P$ we use a 512 atom CuPt structured supercell, [15, 16] with alternating Ga and In atomic layers in the [111] direction. For disordered $Ga_{0.5}In_{0.5}P$ we use 216 atom supercells. (Pair correlations are not optimized, but would have little impact, since the strongest effect turns out to be internal strain - see section 3.) K-point integration is at the $\vec{k} = \frac{1}{4}(1,1,1)$ special point, and the planewave cutoff is 250eV.

Since the two examples defects are both p-type dopants, and we consider the strong doping regime, we assume that the Fermi level lies at the VBE, hence $\varepsilon_F = 0$. For the chemical potentials we use stoichiometric values. If we define $\mu_i = \mu_i^{Bulk} + \Delta\mu_i$, then these correspond to $\Delta\mu_i^{stoich} = \frac{1}{2}\Delta H^{form}$ for $Ga_xIn_{1-x}P$. Here, ΔH^{form} is the formation enthalpy of the compound, such that $\Delta H_{\mathrm{AB}}^{form} = \mu_{\mathrm{A_xB_{1-x}}} - x\mu_{\mathrm{A}}^{Bulk} - (1-x)\mu_{\mathrm{B}}^{Bulk}$ and the denominator of the prefactor is 2 in this alloy. The bulk chemical potentials are μ_{Ga}^{Bulk}=-3.610 eV, μ_{In}^{Bulk}=-3.269 eV, μ_{P}^{Bulk}=-6.028 eV, μ_{Zn}^{Bulk}=-1.891 eV and μ_{Cd}^{Bulk}= -1.531 eV. This gives formation enthalpies of -0.59 eV in CuPt-ordered $Ga_{0.5}In_{0.5}P$, -0.62 eV for disordered $Ga_{0.5}In_{0.5}P$.

3. Results

For undoped disordered $Ga_{0.5}In_{0.5}$, full relaxations of both ions and volume gives a_0=5.606 Å. The CuPt-ordered structure has a symmetry axis [111], so without dopants the Ga-P bonding distances are 2.32 Å parallel to [111] and 2.40 Å perpendicular to it, with the corresponding In-P bonding distances being 2.52 Å and 2.47 Å. The resulting average lattice parameter is a_0=5.614 Å. These a_0 values (5.606 Å and 5.614 Å) differ by only +0.07% and -0.05% from the prediction of Vegard's law, which is 5.61 Å using the LDA values for a_0^{InP} and a_0^{GaP}. On the other hand, the LDA VBE of $In_xGa_{1-x}P$ is not quite so linear. It lies 0.456 eV higher in pure GaP than in pure InP. A linear interpolation therefore predicts a value for $In_{0.5}Ga_{0.5}P$ which is 0.312 eV higher than in pure InP. The values obtained are 0.341 eV and 0.390 eV for the ordered and disordered alloys respectively, which differ by 9% and 25%.

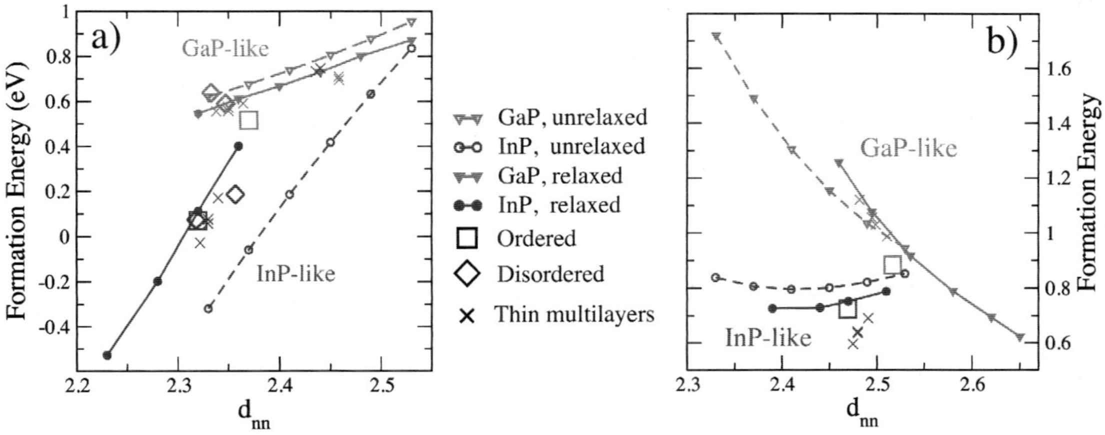

Figure 1. Formation energy of a) Zn^-_{III} and b) Cd^-_{III} in $In_xGa_{1-x}P$, versus average bond length at the dopant site. Lines link values in strained InP or GaP, with (solid) or without (dashed) local relaxation after substitution. (Multilayer data from [7], further details in [17].)

In the ordered alloy, the dopants can occupy two possible sites: in the Ga-planes or in the In-planes. For Zn, E^F_q is 0.45 eV lower when substituting In rather than Ga (see Fig. 1) so the majority of Zn^-_{III} should locate on the In sublattice. Furthermore, E^F_q of Zn^-_{In} is actually 0.35 eV lower than in pure InP, indicating a significant increase in solubility. On the other hand, for Zn^-_{Ga}, E^F_q is practically the same as in pure GaP -just 0.01 eV higher. For cadmium, E^F_q of Cd^-_{In} is close to that in pure InP (0.07 eV lower), but for Cd^-_{Ga} it is 0.38 eV more stable in the ordered alloy than in pure GaP. (In this case the stability improvement of Cd^-_{Ga} will not affect the solubility, as Cd^-_{In} remains 0.16 eV more stable.)

For Zn in the disordered alloy, we consider only sites with either a maximal number of In or of Ga second neighbours, or half of each. As expected, there is a wide spread of E^F_q values, though we only plot the highest two (both with Ga second neighbours) and the lowest two (both with In second neighbours) in Fig 1. We find that the type of second neighbour has rather less influence than the identity of the substituted cation. In practise, only the sites with the lowest formation energies are likely to be occupied, and for these we get very similar results to those in the ordered alloy (a difference of only 0.004 eV). Hence Zn is again considerably more stable than in either pure InP or pure GaP. We have noted elsewhere [7, 17] that we also find a very similar increase in stability in an $(InP)_x/(GaP)_{1-x}$ multilayer structure. (The data are reproduced in Fig 1.) For the multilayers we estimated the resulting defect solubility and its dependence on x. For $x = 0.1$ we found a solubility five orders of magnitude larger than in either pure InP or pure GaP. It seems reasonable to expected a very similar solubility increase if we tune x for either the random or the ordered alloys.

Fig 1 also shows E^F_q for Zn^-_{III} and Cd^-_{III} in pure GaP and InP under strain, plotted both with and without additional ion relaxation after the substitution. Allowing this relaxation lowers E^F_q by an almost constant amount, so the local relaxation energy is more-or-less independent of lattice parameter. Also, when the dopant has a similar size to the host atom substituted (Zn on Ga or Cd on In) there is little local relaxation. If we now compare the relaxed E^F_q in strained pure materials, to those in ordered and disordered alloys and multilayers, we find that they are very similar. This is significant, since it is computationally much cheaper to calculate E_q^F in the strained pure materials than in either the ordered or disordered alloys, and yet this result gives a reasonable estimate (±0.1 eV) of E_q^F in the alloy.

4. Discussion and Conclusion

It is clear that the formation energy of dopants in semiconductor alloys depends strongly on the local structure, and in particular on the local strain. It is also clear that there is very little effect, regardless of the identities of second neighbours, when the dopant is of a similar size to the ion substituted (Zn on Ga or Cd on In). Large increases in dopant stability relative to the pure materials occur only when there is a large difference in ionic radii (Zn on In or Cd on Ga.) For example, when Zn_{In}^- forms in InP, the four Zn-P bonds formed are stretched significantly beyond their equilibrium length, destabilizing the dopant. However, when we alloy with GaP, the lattice parameter is reduced, since Ga is smaller than In. The Zn-P bonds are then allowed to return closer to their preferred lengths, reducing the formation energy and increasing the stability and hence solubility. Similarly, when Cd_{Ga}^- forms in GaP, the resulting Cd-P bonds are over-compressed. Alloying with InP then increases the lattice parameter, releasing this stress and lowering the formation energy. To summarize this, if we wish to increase the solubility of dopant Z in alloy A_xB_{1-x} (where the formation energy is lower in A than in B), the solubility of Z in A can be raised by alloying it with B if either:

1. If Z is smaller than the substituted ion in A and $a_0^B < a_0^A$
2. If Z is larger than the substituted ion in A and $a_0^B > a_0^A$

Since the identity of the substituted ion is important, (rather than that of the second neighbours,) direct experimental studies of this effect would probably depend upon precise preparation methods, with, say, ion implantation being more likely to achieve a solubility increase. Such studies have not been made to our knowledge, but the results of several earlier studies may be attributed to our strain-solubility mechanism (see [7, 17]). For example, Mg_{Ga} accumulates at the interface when cubic GaN is grown on a GaAs substrate.[18] Since Mg atoms are larger than the substituted Ga atoms and GaN is stretched near the interface, the formation energy of Mg_{Ga} should be lowered near the interface, explaining the accumulation. Similarly, in some InP/InGaAsP lasers, Zn dopants diffuse over time, and build up inside the alloyed active region. [19] Our results suggest that Zn would indeed be on average more stable there, though further work is needed to confirm this.

References

[1] W Kohn and L J Sham 1965 *Phys. Rev.* **140** A1133
[2] L Vegard 1921 *Z. Phys.* **5** 17
[3] E G Seebauer and M C Kratzer 2006 *Mat Sci Eng R* **55** 57
[4] C W M Castleton, A Höglund and S Mirbt 2009 *Modelling Simul. Mater. Sci. Eng.* **17** 084003
[5] A Höglund, C W M Castleton, M Göthelid, B Johansson and S Mirbt 2006 *Phys. Rev. B* **74** 075332
[6] L Nordheim 1931 *Ann. Phys.* **9** 607
[7] A Höglund, O Eriksson, C W M Castleton and S Mirbt 2006 *Phys. Rev. Lett.* **100** 105501
[8] A Höglund, C W M Castleton and S Mirbt 2003 *Phys. Rev. B* **72** 195213
[9] C W M Castleton and S Mirbt 2005 *Phys. Rev. B* **68** 085203
[10] J P Perdew and A. Zunger 1981 *Phys. Rev. B* **23** 5048
[11] D Vanderbilt 1990 *Phys. Rev. B* **41** 7892
[12] G Kresse and J Hafner 1994 *J. Phys.: Cond. Matt.* **6** 8245
[13] G Kresse and J Furthmüller 1996 *Comput. Mater. Sci.* **6** 15
[14] M Leslie and M J Gillian 1985 *J.Phys.C: Solid State Phys* **18** 973
[15] A Gomyo, T Suzuki and S Iijima 1988 *Phys.Rev.Lett.* **60** 2645
[16] M Kondow, H Kakibayashi, T Tanaka and S Minagawa 1989 *Phys. Rev. Lett.* **63** 884
[17] A Höglund, O Eriksson, C W M Castleton and S Mirbt 2006 *In Preparation*
[18] D J As *et al.* 1998 *Appl. Phys.Lett.* **73** 1835
[19] T Kallstenius, J Bäckström, U Smith and B Stolz, 1999 *IEEE J. Lightwave Tech.* **17** 2584

Journal of Physics: Conference Series **242** (2010) 012015 doi:10.1088/1742-6596/242/1/012015

Simulation of the tunnelling transport in ferromagnetic GaAs/ZnO heterojunctions

E. Comesaña, M. Aldegunde and A.J. García-Loureiro

Dept. de Electrónica e Computación, Universidade de Santiago de Compostela, 15782 Spain

E-mail: enrique.comesana@usc.es

G.A. Gehring

Department of Physics and Astronomy, University of Sheffield, Sheffield S3 7RH, UK

Abstract. In this work we have implemented a numerical simulator and analytical model to study the dependence of the tunnelling current on the polarization ratio of the carrier spin for a degenerate and ferromagnetic heterojunction. We have applied these models to study the behaviour of a magnetically doped GaAs/ZnO PN junction and the current transport in a PN heterojunction where the polarization of the spin of the charge carriers is also a control variable.

1. Introduction

Spintronic devices based on diluted magnetic semiconductor (DMS) have been studied in the last years [1] due to the simultaneous control of both charge and spin of the carriers. Two of the most popular DMSs are (Ga,Mn)As and (Zn,Co)O, the first one because it is compatible with the current GaAs technology [2] and the second one because of its high Curie temperature [3]. Structures such us PN junctions [1] and resonant tunnelling devices [4] are being fabricated and studied. We develop analytical and numerical models that facilitate the study of magnetic tunnel junctions (MTJ) because of their potential applications as new magnetic sensors or magnetoresistive random access memory (MRAM). In this work we focus on the analysis of a theoretical GaAs/ZnO PN junction doped with magnetic impurities to study the dependence of the charge and spin transport on the spin polarisation of the charge carriers. The analytical model [5] is based on Kane's theory [6] and the numerical simulator solves the drift-diffusion equations modified to consider the spin polarization. These two models allow us to predict the behaviour of the tunnel magnetoresistance (TMR) with the applied bias and temperature variation. In this paper we will show how we calculate the split bands in section 2, the tunnelling current in section 3 and the characteristics of the analytical model and how we implement the numerical model in sections 4 and 5, respectively. Finally we summarize our current results and the conclusions in section 6.

2. Spin splitting

Adding magnetic dopants to the semiconductors leads to the appereance of an exchange interaction between the carriers and the magnetic impurities. The effect on the energy band structure and transport properties can be studied using perturbation theory. It leads to a first order

© 2010 IOP Publishing Ltd

Journal of Physics: Conference Series **242** (2010) 012015 doi:10.1088/1742-6596/242/1/012015

correction in the band energies which means that the bands become spin split [7]. The new energy bands are:

$$E_\sigma = E_0 + \frac{\Delta}{2}\left(\delta_{\sigma\uparrow} - \delta_{\sigma\downarrow}\right) \qquad (1)$$

where E_0 is the energy band when there is no spin polarisation, $\sigma = \{\uparrow, \downarrow\}$ means spin up and down and $\Delta = xJ_{exch}\langle S^z\rangle$ is the energy splitting value, which depends on the magnetic impurity concentration (x), the strength of the exchange interaction (J_{exch}) and the average spin polarization which is obtained using the Brillouin function:

$$x\langle S^z\rangle = xSB_S\left(\frac{g_L\mu_B S B_{eff}}{K_B T}\right) \qquad (2)$$

where S is the net spin that each magnetic impurity contributes ($S = 5/2$ for Mn) and B_{eff} is the effective molecular magnetic field produced by the coupling between magnetic ions and charge carriers. Figure 1 shows the variation of the splitting energy with the temperature in $Ga_{1-x}Mn_xAs$ calculated using the previous model with a manganese concentration of $x = 0.05$ and a carrier concentration of $p = 3\cdot10^{20}$ cm^{-3}. The Curie temperature is 110 K for this carrier concentration and above that temperature the ferromagnetic properties disappear.

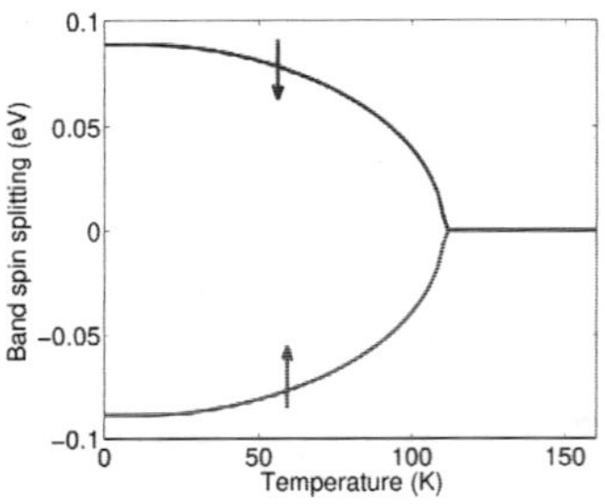

Figure 1. Dependence of band splitting for (Ga,Mn)As on the temperature. The Curie temperature in this case, $p = 3\cdot10^{20}$ cm^{-3}, is 110K. Above that temperature the ferromagnetism disappears.

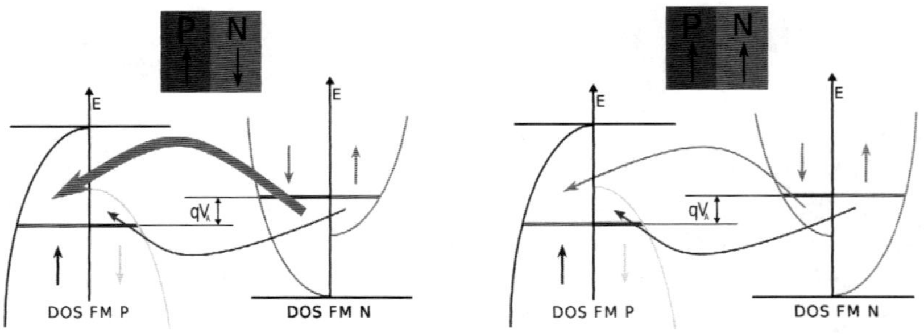

Figure 2. Schematic representation of the spin split bands in a ferromagnetic PN junction. Two different configurations, one for (a) antiparallel and another for (b) parallel magnetisations between both sides of the junction.

3. Tunnelling current

Due to the high carrier concentration required in DMS to ensure ferromagnetic behaviour near or above room temperature, these are highly degenerate semiconductors. Therefore, the carrier transport at low bias will be dominated by tunnelling processes. The probability that an electron tunnels through the depletion region from the conduction band to the valence band is calculated using the WKB approximation:

$$\Gamma_{c\to v} = \exp\left(-2\int_l^u \frac{\kappa_c(x,\epsilon)\cdot\kappa_v(x,\epsilon)}{\sqrt{\kappa_c^2(x,\epsilon)+\kappa_v^2(x,\epsilon)}}dx\right), \qquad (3)$$

where κ_c and κ_v are the wavenumber vectors, which contain information about the effective mass and the energy of the carriers that take part in the tunnel process, and u and l are the classical turning points in the potential barrier.

Figure 2 shows a schematic representation of the two possible configurations in a MTJ for the tunnelling current depending on the relative magnetization between the P and N sides. The one shown in figure 2(a) has a low resistance because the majority spin for holes and electrons are recombining. We call this antiparallel configuration (I_a). However, in figure 2(b) the MTJ is more resistive and we call it parallel configuration (I_p). The TMR is evaluated as the relative difference of the tunnelling current in both cases, $TMR = |I_p - I_a|/(I_p + I_a)$.

The tunnel mechanisms are added to the drift-diffusion model as recombination terms, using a non-local transport model [8]:

$$R_{c\to v}(u,l,\epsilon) = \frac{A_{cv}}{qk_B} \times F \times \Gamma_{c\to v}(r) \times T \times \log\left[\frac{1+\exp\left(\frac{E_{Fp}(u)-\epsilon}{k_BT}\right)}{1+\exp\left(\frac{E_{Fn}(l)-\epsilon}{k_BT}\right)}\right], \tag{4}$$

where the logarithmic factor is a supply function of carriers available to tunnelling, F is the electric field, E_{Fn} and E_{Fp} are the Fermi pseudo-levels for electrons and holes, ϵ is the carrier energy and A_{cv} is the effective Richarson constant.

4. Analytical model

The analytical model solves an abrupt junction using the depletion region approximation. The energy band diagram is calculated considering a triangular barrier plus the discontinuity introduced by the different properties of the materials on each side of the junction. The tunnelling current I_t is calculated considering the overlapped states in the conduction band and the valence band, which are obtained using the parabolic approximation for the density of states:

$$N_c^\sigma \sim \left(E_C - E - \frac{\Delta_n}{2}(\delta_{\sigma\uparrow} - \delta_{\sigma\downarrow})\right)^{1/2} \quad \text{and} \quad N_v^\sigma \sim \left(E_V - E - \frac{\Delta_p}{2}(\delta_{\sigma\uparrow} - \delta_{\sigma\downarrow})\right)^{1/2}, \tag{5}$$

the Fermi statistic to obtain the occupancy of the overlapped states, $F_c(E) - F_v(E)$, and the tunnelling coefficient $\Gamma_{c\to v}$:

$$I_t \propto \sum_{\sigma,\sigma'} \int_{E_{\min}}^{E_{\max}} [F_c(E) - F_v(E)]\Gamma_{c\to v}(E, E_C^\sigma, E_V^{\sigma'})N_c(E, E_C^\sigma)N_v(E, E_V^{\sigma'})dE \tag{6}$$

where $\sigma = \{\uparrow,\downarrow\}$, $\sigma' = \{\downarrow,\uparrow\}$ and Δ_n and Δ_p are the splitting values for the conduction and valence band, respectively.

5. Transport equations in the numerical simulator

The numerical simulator solves the Poisson equation and the continuity equations consistently with the spin split bands calculated in each point of the device. The drift-diffusion equations ($\nabla J_n = -q_e(R^\uparrow_{c\to v} + R^\downarrow_{c\to v})$ and $\nabla J_p = q_e(R^\uparrow_{c\to v} + R^\downarrow_{c\to v})$) are modified to consider the degenerate case using a correction factor $\xi(\eta) = \frac{F_{1/2}(\eta)}{F_{-1/2}(\eta)}$ to take account of the degeneration level:

$$J_n = \mu_n\left[\xi(\eta_n)k_BT\nabla n + n\nabla E_C\right] \quad \text{and} \quad J_p = -\mu_p\left[\xi(\eta_p)k_BT\nabla p - p\nabla E_V\right]. \tag{7}$$

The carrier concentration, $n = n^\uparrow + n^\downarrow$ and $p = p^\uparrow + p^\downarrow$, is calculated using the Fermi integral, $n^\sigma = \frac{N_c}{2}F_{1/2}(\eta_n^\sigma)$ and $p^\sigma = \frac{N_v}{2}F_{1/2}(\eta_p^\sigma)$ and $\eta_n^\sigma = (E_{Fn} - E_C^\sigma)/k_BT$ and $\eta_p^\sigma = (E_{Fp} - E_V^\sigma)/k_BT$.

6. Results

We use the following parameters: $E_g = 1.4$ eV, $\chi = 4.07$ eV, $\epsilon_r = 13.2$, $m_n^* = 0.06m_0$ and $m_p^* = 0.5m_0$ for GaAs and $E_g = 3.4$ eV, $\chi = 4.20$ eV, $\epsilon_r = 8.6$, $m_n^* = 0.24m_0$ and $m_p^* = 0.59m_0$ suitable for ZnO [9].The doping levels are 10^{20} cm^{-3} in the N-side and $3 \cdot 10^{20}$ cm^{-3} in the P-side. The calculations were done considering a 70% of spin polarization and the temperature was set to 77 K.

Figures 3(a) and 3(b) show the energy band diagram obtained from the analytical model and the numerical simulator, respectively. In figure 3(b) it is possible to see how the splitting disappears as the carrier concentration drops in the depletion region. The analytical model does not show this because the splitting is constant.

Figures 3(c) and 3(d) show the tunnelling current and TMR from the analytical model and the numerical simulator, respectively. The red solid line is the tunnelling current in a non ferromagnetic diode and the blue dashed line and the green dot-dashed line are the tunnelling current for the parallel and antiparallel configurations. The I–V characteristic shows that in the region where the diode is showing negative resistance, in biases between 0.2 V and 0.4 V, the TMR reaches the highest values. Half way in the negative resistance region, bias $\sim$ 0.3 V, the analytical model shows a higher value of the TMR, $\sim$ 50%, than in the drift-diffusion model, $\sim$ 40%. This can be explained by the constant splitting in the analytical model that reduces the tunneling-barrier width with respect to the numerical model, giving larger values of the parallel tunnelling current. The TMR for larger biases can not be considered because other components of the current (defect assisted or thermal current) have to be taken into account. The values of the TMR are in the same range as experimental values in resonant MTJs [4].

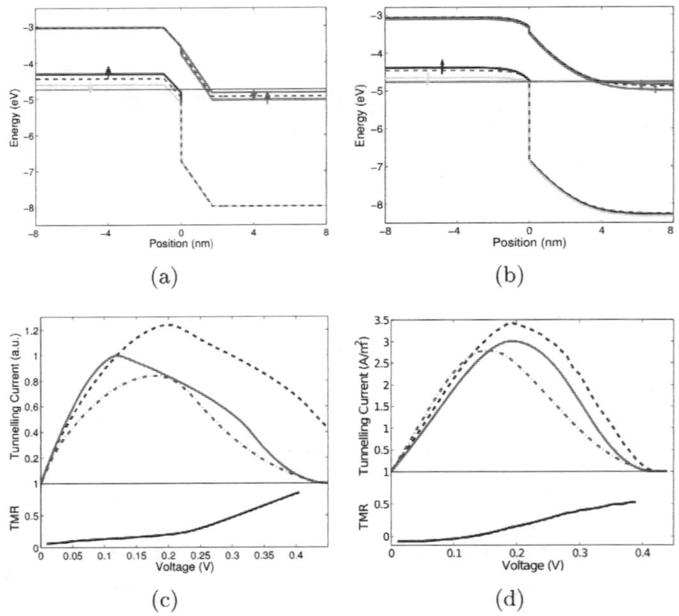

Figure 3. Simulation results from the analytical model (a), (c) and the drift-diffusion simulator (b), (d). Energy band diagram and together the I–V characteristic and the TMR. In both the analytical and numerical case the spin polarisation ratio is 70%.

The analytical model is giving more optimistic results for the TMR than the numerical model. However, the numerical model allows a more precise description of the tunnelling current and its dependence on the spin splitting that leads to more realistic results.

Acknowledgements

This work was supported by Spanish Government (MCYT TIN2007-67537-C03-01) and by Xunta de Galicia (DXIDI07TIC01CT and INCITE08PXIB206094PR).

References

[1] Ohno Y et al. 1999 *Nature* **402** 790–792
[2] Ohno H 1999 *Journal of Magnetism and Magnetic Materials* **200** 110–129
[3] Behan A J et al. 2008 *Physical Review Letters* **100** 0472062008
[4] Chen G, Zeng F and Pan F 2009 *Applied Physics Letters* **95** 232508
[5] Comesaña E and Gehring G A 2007 *Applied Physics Letters* **91** 142510
[6] Kane E O 1961 *Journal of Applied Physics* **32** 83–91
[7] Lebedeva N and Kuivalainen P 2003 *Journal of Applied Physics* **93** 9845–9864
[8] Ieong M et al. 1998 *Proc. Int. Electron Devices Meeting IEDM'98 Tech. Digest* pp 733–736
[9] Pearton S J et al. 2003 *Superlattices and Microstructures* **34** 3–32

Full-Band NEGF simulations of surface roughness in Si Nanowires

A Martinez[1], **A. R. Brown and A. Asenov**
Dept of Electronics & Electrical Engineering, University of Glasgow
Rankine Building, Oakfield Ave, Glasgow G12 8LT, UK

E-mail: antonio@elec.gla.ac.uk

Abstract. In this paper, using the Non-Equilibrium Green Function (NEGF) formalism, we have calculated the electron transport through a silicon nanowire with a rough interface. The nanowire was orientated in the [100] direction. Two different wire diameters, $D_1=2.18\ nm$ and $D_2=3.21\ nm$, have been used to analyse the effect of the cross section in the transmission coefficients. We used a full-band description of the electron state by using a tight binding formalism to describe the electron Hamiltonian. The calculation makes full use of the recursive algorithm to calculate the transmission. We compare several nanowires that differ in the microscopic realization of interface roughness. The energy shift between the onset of the transmission of the rough and the pristine nanowire are *200 meV* and *50 meV* for D_1 and D_2 respectively. Furthermore, the transmissions of some nanowires show resonances due to a resonant cavity created by the interface surface. This is particularly relevant as effective-mass-approximation NEGF simulations of similar nanowire transistors show the same resonances. These resonances play a role in increasing the off-current of nanowire transistors.

1. Introduction

Nanowire (NW) transistors and other 3D devices are expected to replace planar MOSFET technology because they provide better electrostatic control. NWs have been fabricated and their current-voltage characteristics have been measured by attaching metal electrodes at both ends of the NWs [ref 1] However, the roughness of the Si/SiO_2 interface degrades the transport of electrons through the channel. Furthermore, the fluctuations in the cross-section caused by the roughness induce large shifts in the threshold voltage and variability. For NWs under $4\times4\ nm^2$ cross-section the band-structure and electron effective mass are very sensitive to the cross section variation along the wire. We have previously studied the effect of interface roughness on the performance of a gate-all-around Si NW transistor using Non-Equilibrium Green's Functions in an effective mass context [2]. The effective mass approximation (EMA) ignores the valley-splitting effects, which breakdown the degeneration of the lower energy sub-bands of the electrons. Therefore it is important to calculate the magnitude of this splitting and its effect on the transmission. It is well known that the transmission has a direct relation to the electron current through the Landauer formula [3].

Other problems with the EMA are the use of parabolic energy dispersion, which underestimates the density of states at high energies, and other effects caused by non-parabolicity in the *E-k* relation. Nevertheless, the EMA is more computationally efficient due to the spatial representation of the

[1] To whom any correspondence should be addressed.

© 2010 IOP Publishing Ltd

Journal of Physics: Conference Series **242** (2010) 012016 doi:10.1088/1742-6596/242/1/012016

Hamiltonian, which translates in the tight binding context to the use of only one orbital per atom or point. The EMA Hamiltonian can be parameterized to reproduce similar current-voltage characteristics to the TB method. These parameterizations include mass renormalization and introduction of energy dependence in the masses. Therefore it is important to know which details of the band structure still remain after the energy and bands integrations due to the thermal averages.

2. Non-Equilibrium Green Function, The Recursive algorithm and Tight Binding model

The quantum carrier transport is described using the NEGF approach, which is a generalisation of Landauer's formalism [4] for including many body effects. Under certain approximations such as Migdal's theorem [5] (for electron-phonon interaction) and Hartree's approximation [6] (mean field approximation), the NEGF equations can be reduced to a closed equation for the one particle Green's functions. These approximations have been used extensively in the literature to simulate nanoelectronic devices. Here we neglect phonon scattering and consider only scattering and body thickness fluctuation induced by the surface roughness.

The calculation of transmission was carried out by the use of a recursive Green function algorithm described in Reference [7]. We divided the nanowire in layers corresponding to atomic planes perpendicular to the axis of the wire. Because we assume a nearest neighbour interaction, each layer consists of one atomic plane. There are four layers in a unit cell in [100] of a nanowire. The equation for the retarded green function is:

$$\left(E - H_{TB} - V_{Poisson} - \Sigma^{R}_{cont}\right)G^{R} = I \tag{1}$$

where H_{tb} is the tight binding Hamiltonian, $V_{Poisson}$ is the electrostatic potential coming from the Poisson equations and Σ^{R}_{cont} are the contact self-energies representing the source Σ^{R}_{S} and drain Σ^{R}_{D} self-energies. The equation for the calculation of the transmission in this formalism is:

$$T(E) = trace(G^{R}\Gamma_{S}G^{A}\Gamma_{D}) \tag{2}$$

$$\Gamma_{S,D} = -2 \text{ Imag } (\Sigma^{R}_{S,D}) \tag{3}$$

The transmission is related to the current from source to drain by the well-known Landauer expression [3]:

$$I = \frac{q^2}{h} \int T(E)(f_S(E) - f_D(E))dE \tag{4}$$

where f_S and f_D are the Fermi distributions of electrons at the source and drain respectively. This relation shows the importance of the transmission in the current that could be developed in the nanowire when a potential is applied between source and drain.

A ten orbital $sp^3d^5s^*$ tight-binding Hamiltonian [8] with coupling only between the first nearest-neighbours has been used. This model reproduces the bulk electron masses of the electrons with great accuracy.

3. Simulation results

Two nanowires which differ in cross-section radius, $D_1=2.18nm$ and $D_2=3.21nm$, have been considered. The nanowire is orientated in the [100] direction. The length of the region with the rough interface is *6.6 nm*. The interface is introduced by adding/removing an atomic layer at the interface. Then the outer atoms are passivated by hydrogen to avoid states inside the gap. Figure 1 shows the atomically resolved nanowire with diameter D_2. This structure is used to calculate the 1D conduction band structure, which is show in Figure 2. The right panel shows a close-up of the bottom of the conduction band, where the split of the degeneration can be observed; in this case the energy shift is around 12 meV. The corresponding shift for the D_1 diameter nanowire is approximately 24 meV. One of the rough wires (D_2) is shown in Figure 3. Cavities can be observed in this structure as a consequence of the body thickness variation.

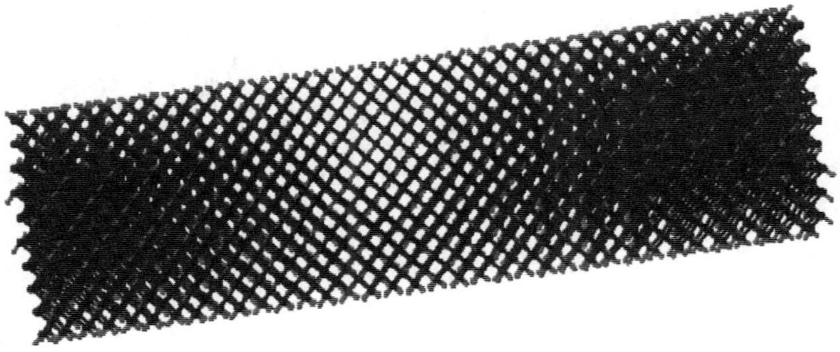

Figure 1. Pristine Si nanowire with diameter D_2=*3.21nm* passivated by hydrogen.

The transmissions for nanowires of two different diameters are shown in Figures 4 and 5 for diameter D_1 and D_2 respectively.

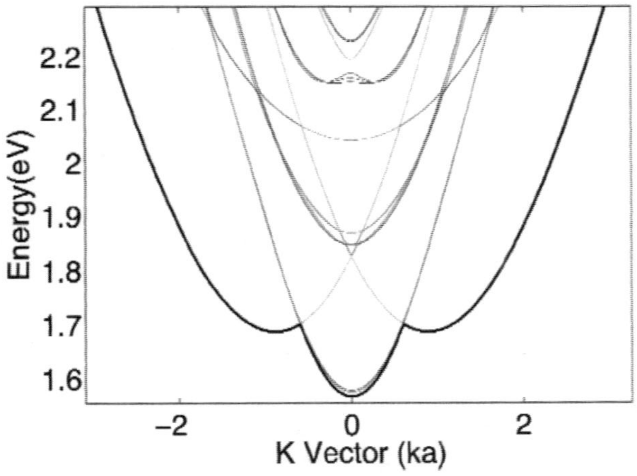

(a)

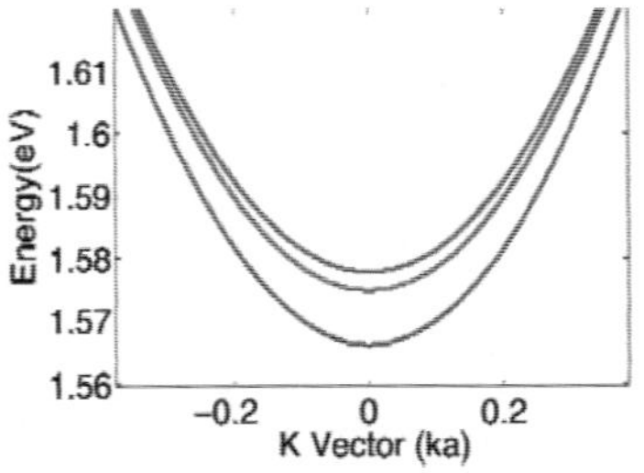

(b)

Figure 2. (a): 1D conduction band structure corresponding to the nanowire of Figure 1. (b): A close-up of the bottom of the conduction band (*a* is the unit cell size) showing the absence of the degeneration.

There is a visible shift of *200 meV* between the onset of transmission in the rough device and the pristine (without roughness) device. Furthermore, a resonant peak appears at around *1.9 eV*. Resonances in transmission have been observed previously in the effective mass model [2] Resonances play an important role in the device performance and are a manifestation of tunnelling through cavities created by the interface roughness. Interface roughness will not always induce these resonances. Furthermore, the existence of the resonant transmission is related to the channel length/correlation ratio. When this ratio is much larger than one, the existence of many aligned random distributed cavities make the creation of resonances in the transmission difficult.

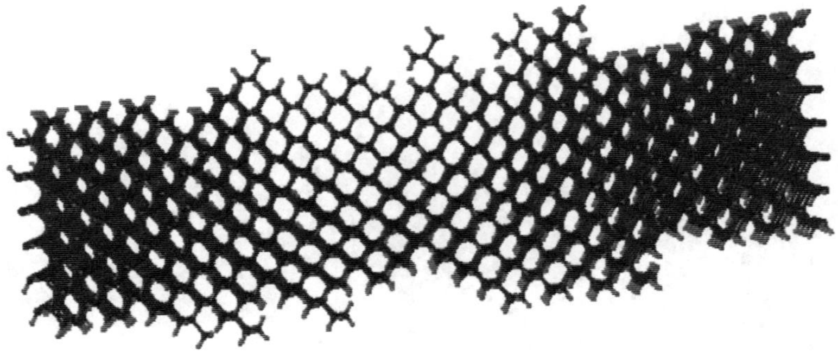

Figure 3. Silicon Nanowire with rough interface passivated by hydrogen with diameter D_2=3.21nm

The transmission coefficients for the large diameter rough nanowires show less effect of the interface roughness. The shift between the onset of transmission in the rough device and the pristine device is *50 meV*. Furthermore, a small resonance is observed around *1.63 eV*. Note, there is an obvious correlation between the band structure in Figure 2 and the steps in the pristine nanowire transmission, e.g. the 2 modes appearing at *1.7 eV* comes from the bulk Δ_2 valleys.

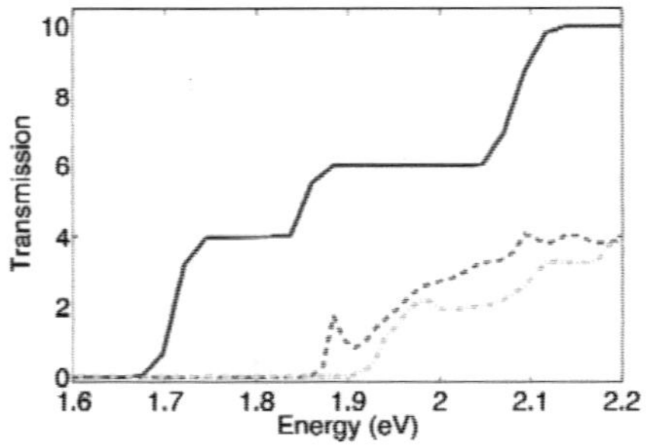

Figure 4. Transmission coefficients for a pristine wire (full line) and for two rough wires. The diameter is D_1.

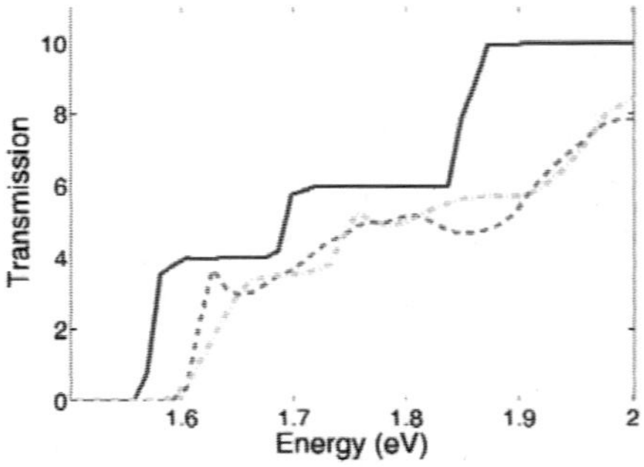

Figure 5. Transmission coefficients for a pristine wire (full line) and for two rough wires. The diameter is D_2.

4. References

[1] Wang J F, Gudiksen M S, Duan X F, Y. Cui and Lieber C. M. 2001 *Science* **293** 1289
[2] Martinez A, Seoane N, Brown A, Barker J R and Asenov A 2010 Submitted to *IEEE Electron. Dev.*
[3] Meir Y and Wingreen N S 1992 *Phys. Rev. Lett.* **68** 2512,
[4] Datta S 2005 *Quantum Transport, Atom to Transistor* (Cambridge University Press),.
[5] Migdal A B, Translation: 1958 *Soviet physics-JEPT*, **7**, 996.
[6] Inkson J C 1984 *Many-Body Theory of Solids* (Plenum Press)
[7] Svizhenko A, Anantram M P, Govindan T R, Biegel B and Venugopal R 2002 *J. Apps. Phys* **91** 2343.
[8] Jancu J M, Scholz R, Beltram F, and Bassani F 1998 *Phys. Rev. B* **57** 6493

Journal of Physics: Conference Series **242** (2010) 012017 doi:10.1088/1742-6596/242/1/012017

Monte Carlo simulation study of the impact of strain and substrate orientation on hole mobility in Germanium

Craig Riddet, Jeremy R. Watling, KahHou Chan and Asen Asenov

Device Modelling Group, Department of Electronics and Electrical Engineering, University of Glasgow, Glasgow, G12 8LT, Scotland, UK

c.riddet@elec.gla.ac.uk

Abstract. The use of alternative channel materials to maintain device performance with scaling for CMOS technology is an active area of research, with Germanium offering an extremely attractive possibility for pMOSFETs in CMOS. In this paper we use full band Monte Carlo transport simulations to investigate the impact of substrate orientation and biaxial strain on hole mobility in bulk Germanium helping to establish a preferential substrate channel orientation that can maximize carrier mobility for these devices.

1. Introduction

Performance boosters such as process induced strain and the introduction of high-κ/metal gate stacks that help CMOS technology to deliver continuous performance improvement [1] may yield diminishing returns with future scaling. The investigation of Germanium (Ge) as an alternative channel material and potential technology booster has experienced a revived impetus, especially for p-channel MOSFETs [2] where higher carrier mobility and drive current compared to conventional CMOS devices have already been successfully demonstrated [3]. This has so far eluded Ge n-channel development, with a number of possible reasons proposed [3,4], though recent progress has been reported [5]. The already superior hole mobility of Ge as compared to Silicon (Si) can be improved further by employing strain and exploiting different substrate and channel orientations.

2. Simulation Methodology

Here we investigate the Ge hole mobility enhancement due to biaxial strain from a relaxed $Si_{1-x}Ge_x$ buffer layer, with three major substrate orientations: (001), (110) and (111), along with a variety of channel orientations. The fraction of Si in the substrate varies from zero (a Ge substrate) up to 0.6 in steps of 0.1. Fractions of Si above 0.6 are not considered, as they would prove problematic to fabricate with the critical thickness becoming impractical for use. The relationship of applied strain, $\varepsilon_{\parallel}$, to Ge content in the buffer, x, is given by $\varepsilon_{\parallel} = a_{Ge}/a_{Si_{1-x}Ge_x} - 1$ where a_{Ge} and $a_{Si_{1-x}Ge_x}$ are the lattice constants for the channel and substrate respectively. The relevant values for this study are given in Table 1. In all cases the material is assumed to be undoped.

© 2010 IOP Publishing Ltd

Table 1. Relationship between Ge content of the SiGe layer and the applied strain.

Substrate	**Strain**
Ge	0%
$Si_{0.1}Ge_{0.9}$	0.44%
$Si_{0.2}Ge_{0.8}$	0.89%
$Si_{0.3}Ge_{0.7}$	1.33%
$Si_{0.4}Ge_{0.6}$	1.74%
$Si_{0.5}Ge_{0.5}$	2.17%
$Si_{0.6}Ge_{0.4}$	2.59%

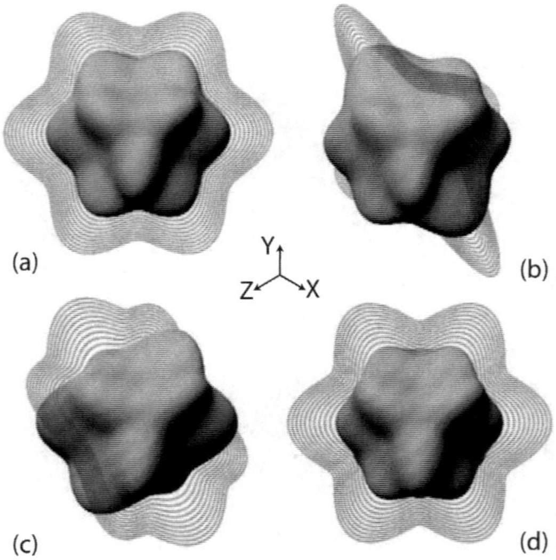

Figure 1. Iso-energy surfaces for the HH band at 1 eV for (a) relaxed Ge and strain on the (b) (001) (c) (110) and (d) (111) substrates. In all cases the surface is viewed from the ⟨111⟩ direction. In-plane energy contours are also shown.

To conduct our study we have employed an ensemble Monte Carlo (MC) technique, encompassing all the major scattering mechanisms [6] though only scattering from acoustic (longitudinal and transverse modes) and optical phonons is included for these bulk-undoped calculations, with parameters calibrated to match the experimentally measured low field mobility and velocity at 220K and 300K. The full bandstructure is obtained using 6-band k•p calculation [7]. Particular care is taken to ensure that the applied electric field in the case of mobility simulations is small enough to avoid undesirable carrier heating [8].

The strain modification of the bandstructure plays an important part in the investigation of the hole mobility in these conditions. The strain removes the underlying 3-fold symmetry and changes the orientation of the iso-energy surfaces [9], as shown in Figure 1 and Figure 2.

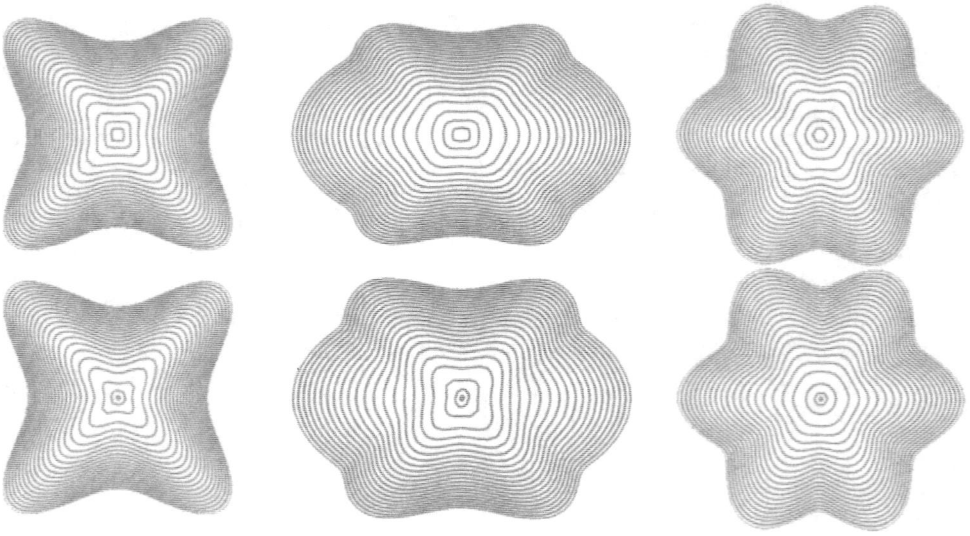

Figure 2. Energy contours for, left to right, (001), (110) and (111) substrates with the relaxed cases shown above the maximally strained versions (Ge on $Si_{0.6}Ge_{0.4}$ substrate). The energies range from 10 meV in the centre out to 2 eV.

This also strongly affects the effective mass and the density of states (DoS), and these are two of the fundamental mechanisms that lead to mobility enhancement. A spherically averaged DoS mass is used in the calculation of the scattering rates, as in [10], and this is plotted in Figure 3 for the heavy (HH) and light hole (LH) bands and for each substrate orientation considered here at 40 meV. The DoS mass progressive decreases for the HH band and levels out towards the maximum level of strain applied in this study. The other significant effect associated with the application of strain to the Ge layer is the energy separation between the normally degenerate HH and LH bands, as illustrated in Figure 4. The consequence of this is that the interband scattering from both the phonon mechanisms is greatly reduced which further leads to a reduction in the occupancy of the light hole band.

While these aspects explain the overall trend of mobility enhancement with strain, they provide little indication of orientation dependence, especially with respect to the channel.

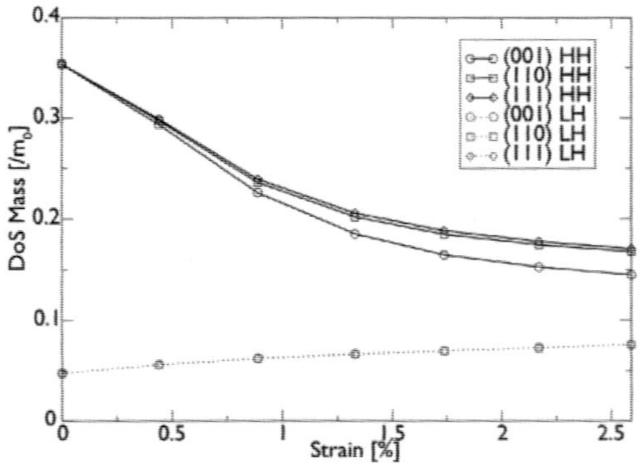

Figure 3. Variation of DoS mass in the HH and LH bands for all 3 substrates in this study.

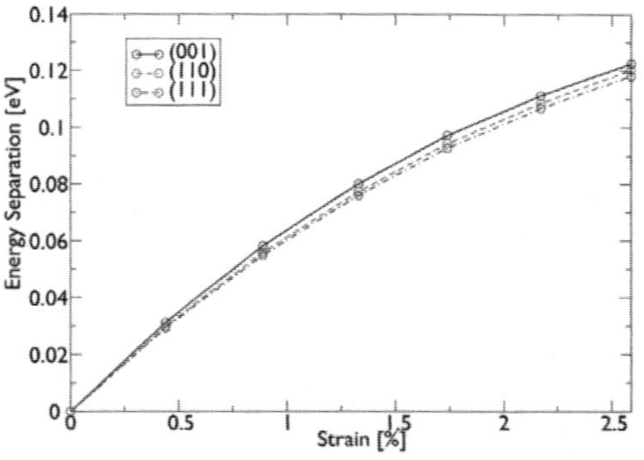

Figure 4. Variation with strain of the energy separation between the HH and LH bands.

3. Results and Discussion

Figure 5 summarises the results of our simulations – only two channel directions were selected for the (001) substrate orientation, as the ⟨010⟩ direction is identical to the ⟨100⟩ direction. We found that the best potential channel orientation is a ⟨-110⟩ orientation on a (110) substrate in terms of mobility. This is consistent with a previously published study, concerning double gate Ge pMOSFETs [11] where only one level of strain is considered. A significant anisotropy is also observed for the (110) substrate that is not present for the other two.

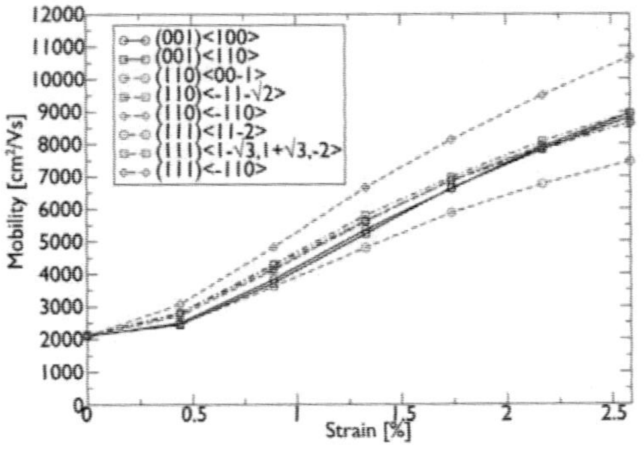

Figure 5. Variation of hole mobility with strain for different substrate/channel orientations.

The energy dispersion relations for the (110) substrate orientation are shown in Figure 6 and provide the best indication of the cause of this directional preference as the strain most significantly impacts upon the shape of the *E-k* relationship for the ⟨-110⟩ (110) channel-substrate direction.

As devices approach the ballistic limit, an important issue for determining device performance is the injection velocity from the virtual-source end of the channel and the corresponding backscattering

Journal of Physics: Conference Series **242** (2010) 012017 doi:10.1088/1742-6596/242/1/012017

coefficient [12]. These issues are currently being investigated. Here special care must be taken to correctly model the ionized impurity scattering, including the effects of the dynamic screening, which plays an increasingly important role for holes. Appropriate modeling of interface roughness scattering is additionally an area of interest for hole transport in Ge inversion layers.

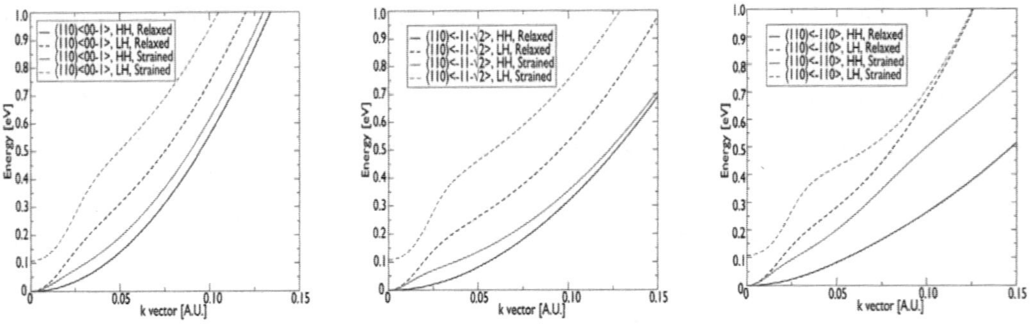

Figure 6. Energy-momentum dispersion relations for the 3 channel orientations on the (110) substrate. The ⟨-110⟩ direction shows the most dramatic reaction to the application of strain that accompanies the superior mobility enhancement observed for this orientation.

4. Conclusions

We have studied the impact of orientation and biaxial strain on hole mobility in bulk Ge using full-band MC transport simulations calibrated to experimental data. The change of effective mass and the separation in energy between the HH and LH bands resulting from the application of strain causes a continual mobility increase irrespective of orientation. However, we find that a ⟨-110⟩ channel direction on a (110) substrate offers the greatest increase in overall mobility due to the dramatic impact that strain has on the energy-momentum relationship. Further considerations necessary to model scaled pMOS devices are also presented.

References

[1] Watling J R, Brown A R, Ferrari G, Barker J R, Bersuker G, Zeitzoff P, and Asenov A 2008 *J. Comp. Theoret. Nanoscience* **5** 1072–88

[2] Nicholas G, Jaeger B D, Brunco D P, Zimmerman P, Eneman G, Martens K, Meuris M, and Heyns M M 2007 *IEEE Trans. El. Dev.* **54** 2503–11
Whang S J, Lee S J, Gao F, Wu N, Zhu C X, Pan J S, Tang L J, and Kwong D L 2004 *IEDM Tech. Dig.* 307-310

[3] Kuzum D, Pethe A, Krishnamohan T, Oshima Y, Sun Y, McVittie J P, Pianetta P A, McIntyre P C, and Saraswat K C 2007 *IEDM Tech. Dig.* 723–726

[4] Lieten R R, Degroote S, Kuijk M, and Borghs G 2008 *Appl. Phys. Lett.* **92** 022106
Tsipas P and Dimoulas A 2009 *Appl. Phys. Lett.* **94** 012114

[5] Kuzum D, Krishnamohan T, Nainani A, Sun Y, Pianetta P A, Wong H S-P, and Saraswat K C 2009 *IEDM Tech. Dig.* 453-6

[6] Wiley J D 1971 *Phys. Rev. B* **4** 2485-93

[7] Dijkstra J E and Wenckebach W T 1997 *J. Appl. Phys.* **81** 1259–63
Dresselhaus G, Kip A F, and Kittel C 1955 *Phys. Rev.* **98** 368–384

[8] Ottaviani G, Canali F, Nava F, and Mayer J W 1973 J. Appl. Phys. 44 2917–18

[9] Rideau D, Feraille M, Ciampolini L, Minondo M, Tavernier C, and Jaouen H 2006 *Phys. Rev. B* **74** 1-20

[10] Fischetti M V and Laux S E 1996 *J. Appl. Phys.* **80** 2234–52

[11] Krishnamohan T, Kim D, Dinh T V, Pham A, Meinerzhagen B, Jungemann C, and Saraswat K 2008 *IEDM Tech. Dig.* 899-902

[12] Lundstrom M S 1997 *IEEE Elec. Dev. Let.* **18** 361-3
Jeong C, Antoniadis D A and Lundstrom M S 2009 *IEEE Trans. El. Dev.* **56** 2762-69

AUTHOR INDEX

Institute of Physics
Dirac House, Temple Back
Bristol BS1 6BE UK

CURRAN ASSOCIATES INC
proceedings.com